THE ILLUSTRATED ENCYCLOPEDIA OF WILDLIFE

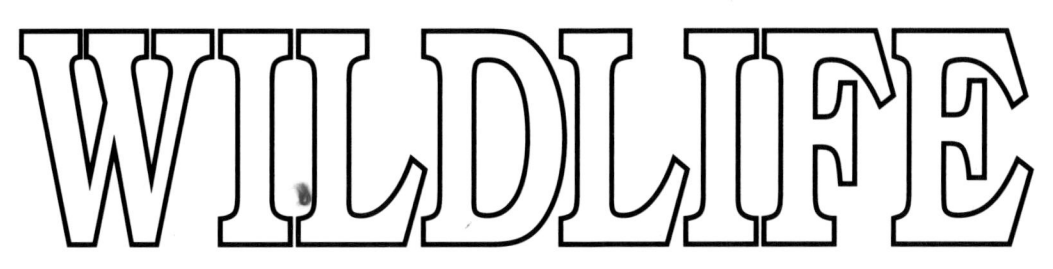

VOLUME 4

The Mammals

Part IV

Wildlife Consultant

MARY CORLISS PEARL, Ph. D.

Distributed by Encyclopaedia Britannica
Educational Corporation

Grey Castle Press

Published by Grey Castle Press, 1991

Distributed by Encyclopaedia Britannica Educational Corporation, 1991

All rights reserved. No part of this book may be reproduced or transmitted in any form or by any means electronic or mechanical, including photocopying, recording or by any information storage and retrieval system, without permission in writing from the Proprietor.

THE ILLUSTRATED ENCYCLOPEDIA OF WILDLIFE
Volume 4: THE MAMMALS—Part IV

Copyright © EPIDEM-Istituto Geographico De Agostini S.p.A., Novara, Italy

Copyright © Orbis Publishing Ltd., London, England 1988/89

Organization and Americanization © Grey Castle Press, Inc., Lakeville, CT 06039

Library of Congress Cataloging-in-Publication Data
The Illustrated encyclopedia of wildlife.
   p.   cm.
   Contents: v. 1–5. The mammals—v. 6–8. The birds —
v. 9. Reptiles and amphibians — v. 10. The fishes —
v. 11–14. The invertebrates — v. 15. The invertebrates
and index.
   ISBN 1–55905–052–7
   1. Zoology.
QL45.2.I44 1991   90–3750
591—dc20   CIP

ISBN 1–55905–052–7 (complete set)
     1–55905–040–3 (Volume 4)

Printed in Spain

Photo Credits
Photographs were supplied by *Ardea London Ltd.*: (F. Gohier) 571, 579t, 583; *Bruce Coleman*: 557, 598t, 606, 622, 623, 627, 630, 701b; (J. & D. Bartlett) 576, 677; (S.C. Bisserot) 679, 683, 688, 691t, 693, 696t, 703, 705l, 705r, 706, 707, 708, 713t, 714r, 715b; (J. & S. Brownlie) 572; (J.R. Brownlie) 633b; (J. Burton) 536t, 536b, 567b, 594, 613r, 615, 649, 650b, 653t, 663, 666, 669, 682, 684, 698; (B.J. Coates) 603; (E. Crichton) 558; (N. Devors) 539; (F. Erize) 548, 55, 590–591, 592m; (J. Foott) 537; (C.B. & D.W. Frith) 562t, 642t, 673, 676; (U. Hirsch) 533, 697b; (D. Houston) 652; (J.L. Hout) 549; (G. Laycock) 586t; (J. Markham) 562b, 604; (N. Myers) 720l; (O. Newman) 661t, 664, 665, 667t; (G. Pizzey) 621, 629t, 643; (M.P. Price) 648b; (H. Reinhard) 546, 547, 553, 560, 645, 659, 670; (L. Lee Rue) 567t, 611t, 612, 613l, 614b; (L. Lee Rue III) 566; (W.E. Ruth) 551; (V. Serventy) 626, 632, 634; (J. Simon) 614t; (N. Tomalin) 545, 660; (J. Van Wormer) 564, 565; (P. Ward) 675; (J. Wallis) 631, 636, 637t; (R. Williams) 559; (G. Ziesler) 704; *I Dominicus*: 575b; *Jacana*: 598b, 600, 617b, 618l, 628l, 629b; (M. Balleau) 589; (A. Bertrand) 618r, 625l; (Bailleau) 658r; (J.P. Champroux) 680; (M. Danegger) 648t; (Devez-CNRS) 654t, 654b; (J-L.S. Dubois) 656; (Gens) 544; (J.P. Hervy) 655, 694, 699, 702; (C. de Klemm) 651b, 695; (Mammifrance) 661b, 667m, 668, 672t, 681, 689, 691b, 709t, 709b, 712t, 712b, 714l; (C. Pissavini) 541; (A. Rainon) 563; (B. Rebouleau) 667b; (J. Robert) 554, 715t; (J.X. Sundance) 671t, 671b, 672b, 674; (R. Tercafs) 653b; (J.P. Varin) 573, 586b, 595l, 595b, 596, 597, 599, 601, 607, 610, 617t, 619t, 620, 625r, 628r, 633t, 635, 637b, 641l, 641r, 642b, 644, 685, 696b, 697t, 700, 710, 711t, 711b; (Varin-Visage) 540, 561, 584, 585, 593t, 592b, 640, 647, 650t, 651t, 658, 662t, 662b, 711m; (Verzier) 552; (R. Volot) 657b, 692l, 692r; (Ziesler) 611b; *S. Lovari*: 542l, 542r, 543; *NHPA*: (A. Bannister) 701t; (S. Dalton) 713b; (P. Fagot) 638, 639; *NHPA/ANT*: (G.E. Schmida) 609; *Oxford Scientific Films*: (M. Conte) 587; (B. Osborn) 582; *Researchers*: (R. Kinne) 557; (R. Sternach) 581bl; *Seaphot*: (IOS) 578, 580; (L. Madin) 579b; (N. Merrett) 575t; (R. Salm) 588l, 588r; *Zefa*: (McCutcheon) 581t.

FRONT COVER: Killer whales (Ardea London Ltd./Ian Beanes)

CONTENTS

Artiodactyla (Even-Toed Ungulates)
    The Bovids
        The Goat Antelopes—*Sure-Footed Climbers*   533

Whales and Dolphins—*Titans of the Deep*   571

Platypus and Echidnas—*Egg-Laying Oddities*   597

Marsupials—*Mammals With Pouches*   607

Insectivores and Elephant Shrews—*The Insect Hunters*   645

The Colugo Family—*Treetop Gliders*   673

The Bats—*Navigators of the Night*   677

THE GOAT ANTELOPES—CAPRINS

SURE-FOOTED CLIMBERS

The goat antelopes thrive in terrain that few other mammals can master, from treacherous cliffs and crags to frozen Arctic tundra

THE GOAT ANTELOPES—CAPRINS

Walia ibex

Takin

Blue sheep

Markhor

Barbary sheep

Chiru

Nilgiri tahr

Himalayan tahr

Mainland serow

THE GOAT ANTELOPES—CAPRINS

The goat antelopes comprise 26 species living in habitats as diverse as deserts, alpine slopes and Arctic wastes. The animals include such famed rock climbers as the wild goat and mouflon—ancestors of the domestic goats and sheep. The group also includes the musk ox, which is superficially more like a bison than a goat. The animals have a wide variety of body shapes, colors and sizes, but they are usually stocky creatures with horns that range from small (in the goral) to massive (in the ibex).

Swollen-nosed grazers

The saiga and the chiru are central Asian animals that form a link between the gazelles and the goats. Their most distinctive feature is a large, curiously shaped swollen nose.

The chiru or Tibetan antelope measures up to 35 in. at the withers and weighs some 65 lbs. With its long legs and long, lyre-shaped black horns, the chiru is superficially similar to a gazelle. But it is more sturdily built and has a short, fleshy protuberance from the nasal bone, almost like an attempt at a trunk, that opens out into two nostrils at the end. During the rut, the male inflates two nasal sacs (lined with mucous membranes) as part of his mating display. Only males bear horns, but both sexes have numerous scent glands. Chirus are yellowish-white with darker areas on the head and flanks, pale undersides and large pale rump flashes. Their coats are bristly, with tufts of long hair over their short tails.

The chiru's stronghold is the high Tibetan plateau, 13,000-20,000 ft. above sea level. During the summer it may climb even higher to take advantage of the seasonal alpine pasture, but in general, it is found in the river valleys where grass is plentiful. Grazing in small groups, chirus stay out on the pasture until sunset. As dusk falls and the temperature drops, they move into cover. Each chiru finds a spot that is sheltered from the harsh weather and digs out a shallow hole where it can rest for the night.

Immune to cold

Chirus are resistant to the severe weather conditions at these altitudes. During the mating season in late autumn, the nighttime temperature can fall to $-22°F$. In spite of this, the males eat very little, even though they use up a huge amount of energy defending their territories, pursuing females and mating.

Adult males are territorial during the mating season. The females wander from one male territory to another within their regular home range. One, or more rarely two, offspring are born outside the herd between May and July after a gestation of about six months. The young animals immediately follow their mothers back to the herd.

PAGE 533 A herd of chamois keep watch from a clifftop vantage point in the Abruzzo National Park in central Italy. They have sharp eyesight and hearing, making them difficult to approach. A chamois can leap up rocks in 13-foot bounds.

GOAT ANTELOPES CLASSIFICATION: 1

The 26 species of goat antelopes are grouped into four tribes: the saiga and chiru form the tribe Saigini; the serows, goral, chamois and mountain goat comprise the tribe Rupicaprini; the takin and musk ox form the tribe Ovibonini; and the remaining 17 species, including the tahrs, the wild goat, the ibex and the mouflon, make up the tribe Caprini. The Rupicaprini, the Ovibonini and the Caprini all belong to the subfamily Caprinae—part of the family Bovidae. The Saigini have also been classified with the Caprinae, but many zoologists now consider them to be linked to the subfamily Antilopinae.

Saigini

The Saigini are a small tribe of ungulates with characteristics midway between those of goats and gazelles. Only the males possess horns. There are only two species within the tribe. The chiru or Tibetan antelope, *Pantholops hodgsoni*, is a mountain animal found in northern India, the Tibetan plateau and central China. The saiga, *Saiga tatarica*, lives on the cold, arid plains of central Asia, from the Black Sea to Mongolia and western China (Sinkiang province). The Mongolian subspecies *S. t. mongolia* is now extremely rare.

THE GOAT ANTELOPES—CAPRINS

The chiru is so specialized for life at high altitudes that it has problems adjusting to conditions at lower levels, and it does not thrive in zoos or wildlife parks. With few chiru in captivity and their natural habitat hard to reach, the chiru has been little studied, and there are still many gaps in our knowledge of its way of life.

The remarkable saiga

The saiga is a close relative of the chiru, and its nasal equipment is even more elaborate. The saiga's nose is enlarged into a hump on top of its snout; a breeding male can inflate the nose beyond its normal puffy shape, giving it a grotesque, swollen appearance that is accentuated by the downward-facing nostrils at the end of its muzzle.

Adult male specimens of the Russian saiga reach a height of about 30 in. at the withers; females are smaller. In summer they have yellowish coats with broad, white throat patches and darker facial markings, but in winter their color changes to near-white to make them less noticeable in the snow. The coat, which is heavy and woolly, is even thicker in winter to combat the cold. Only the males bear horns. These are translucent, densely ringed and rather stumpy compared with those of the chiru, and rarely grow longer than 12 in. The very rare Mongolian subspecies is smaller and brownish in color with a brown patch on its hindquarters.

During the summer, saigas feed on the arid steppelands and semi-deserts of central Asia; in winter they move to hilly country, where they find better shelter from the snow and the sub-zero winds. As a lowland animal, the saiga does reasonably well in captivity, although there are few specimens in zoos and parks outside the Soviet Union.

TOP **Almost wiped out by hunting, the saiga has once again become numerous on the steppes of central Asia.** ABOVE **The saiga's curious swollen nose gives it a distinctive profile. During the rut, the swelling is inflated as part of the male's sexual display.** ABOVE RIGHT **The normal ambling gait of the saiga (A) contrasts with occasional bursts of action; female saiga leap into the air to drive off large birds of prey (B)—if an eagle tries to carry away a young saiga, the mother may jump into the air and attempt to strike the bird with its sharp front hooves, driving it off as it swoops down.**

THE GOAT ANTELOPES—CAPRINS

Natural nomads

Saiga herds often travel many miles each day in search of good grazing, and during migrations they can cover 50-75 miles a day. Saigas are gregarious, and in summer the average herd contains some 30-40 individuals. Sometimes, environmental conditions cause several herds to unite, forming a great herd numbering up to a thousand animals. They stay on the move throughout the year until the mating season in December, when they settle down for a time to allow adult males to gather their harems. These contain 5 to 15 females within territories of 1 to 4 square miles, which the male defends vigorously against rival males. Mating takes place at night, which is unusual for ungulates.

The territory holders are highly aggressive during this period, and the combats that take place are bloody and sometimes fatal for one of the contenders. The males become so aroused that they are unable to graze, although they occasionally eat a little snow, perhaps to soothe their parched mouths. Obsessed with defending their harems against other males, they will attack anything or anyone that comes too close— including humans. Meanwhile, the younger males form separate single-sex groups, taking care to keep well clear of their unpredictable elders, who are liable to attack without the slightest hesitation.

By the end of the mating season, the breeding males are so exhausted that they are too weak to cope with the rigors of a Siberian winter. It is not unusual for over 85 percent of the adult males to die before spring, and in a particularly severe season the survival rate may be less than 5 percent. In late April and May the pregnant females withdraw to give birth alone; single births are usual, but twins are not uncommon. After two days the young saigas can run faster than a man and are well equipped to flee from predators. The females reach maturity at seven to eight months, whereas males become sexually active at about two years. Population density and other environmental factors, however, can also influence sexual maturity.

ABOVE LEFT For a few days after they are born, young saigas lie low among ground-squirrel burrows and anthills in their grassland habitat.
ABOVE Although massively built, the mountain goat of North America is surprisingly agile and sure-footed, spending most of its time clambering among the steep cliffs and rugged crags of its Rocky Mountains home. Insulated by its thick woolly coat, it can survive the freezing sub-Arctic conditions of an Alaskan winter by living on the mosses, lichens and young trees that lie beneath the deep snow.

Back from the brink

The saiga was once commonly found from European Russia to Alaska, but unrestrained hunting took its toll and by the early 1920s it was on the edge of extinction. The Soviet government recognized the danger and placed it under strict protection, and its numbers began to increase. Within a few years the

THE GOAT ANTELOPES—CAPRINS

saiga population had expanded enormously, and from a few hundred animals it rose to several hundred thousand. By 1958 there were an estimated two million saigas in the USSR. The Mongolian subspecies, however, has not shared in this recovery and is now seriously endangered: in the late 1970s the total population was only about 200 animals.

Since their recovery, saigas have been hunted— under strict controls— in the Soviet Union. Apart from man, the saiga's main enemy is the wolf. Hunting in packs, the wolves concentrate on sick or weak animals, killing many of the exhausted males in winter. Young, vigorous saigas are usually able to escape. Eagles, foxes and ravens may take a young animal, especially right after it is born, but the rapid development of the young saigas and the fierce protection provided by their mothers foil many of these attacks.

A mountain life

The goral is the smallest of the goat antelopes. It measures 30-50 in. head-to-tail, standing only 24-30 in. at the withers and weighing 55-75 lbs. There are two species: the common goral and the red goral. The males and females are similar, with short, tapered horns. The coat of the common goral varies in color from gray to reddish brown with a white throat patch, a black stripe along each flank and a small, spiky mane on the neck. The red goral's coat is redder and has no white throat patch. Some subspecies have white feet. The similarity of size between the sexes results from their solitary way of life; there is no advantage for the male in being larger, since it has no herd to defend.

The common goral is found from the southeastern edge of Siberia to northern Burma, northern Thailand and the Himalayas; the red goral lives in Tibet, northern Burma and Assam. Both species live in the mountains at altitudes of 3200-13,000 ft. or more. They prefer rocky slopes, however steep, and ideally covered with medium to high-stemmed vegetation, but they can also adapt well to bare, arid ground. They feed on twigs and leaves of shrubs, grasses and also nuts. A goral will find an area that has a good supply of such foods and stay there; in most cases it rarely strays beyond a restricted home range of about one mile in diameter. In winter it may limit its activities to even smaller areas, provided they are steep enough for fresh snowfalls to slip away easily, leaving the underlying vegetation exposed.

A goral can get all the food it needs from a small area only if there is no competition from other gorals.

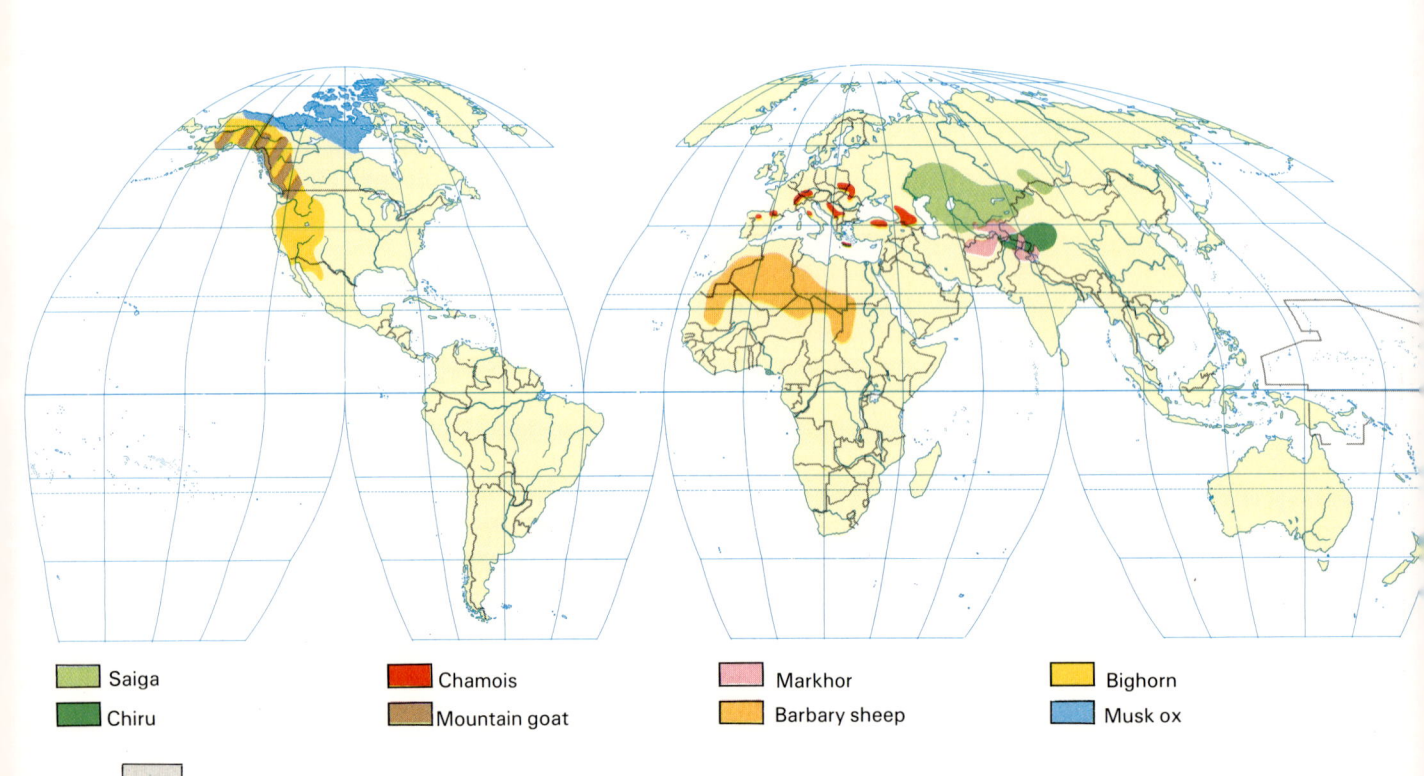

☐ Saiga ☐ Chamois ☐ Markhor ☐ Bighorn
☐ Chiru ☐ Mountain goat ☐ Barbary sheep ☐ Musk ox

THE GOAT ANTELOPES—CAPRINS

Accordingly, it will defend its patch vigorously against other gorals, using its short, sharp horns to murderous effect—the goral and its close relatives display little of the restraint shown by the larger-horned ungulates. They relax this vigilance only during the mating season, when they may gather together into small groups of two to eight individuals.

The serows

The two species of serow are adapted to two extremes of climate. The mainland serow tolerates a warm, humid environment. It lives in the tropical and subtropical forests of the Malay Peninsula, Sumatra, Southeast Asia, Thailand, Burma, central and southern China, Nepal, Assam and Kashmir, in wooded gorges and on mountain slopes. In contrast, the Japanese serow lives in the cold, snowy regions of Japan; a dwarf form also occurs on the island of Taiwan.

The mainland serow is variable in color, but is usually dark brown or dark gray with paler feet,

ABOVE Gleaming white against the dark rock, a mountain goat makes a striking target after the snows have melted. Better roads have allowed more hunters to enter the goats' rough habitat, and increasing numbers of the animals are shot during the hunting season.
ABOVE LEFT Two male mountain goats threaten one another by standing next to each other head-to-tail (top) and indicate submission and dominance (center). Only if all else fails will they resort to using their horns in combat (bottom).
FAR LEFT The map shows the geographical distribution of several species within the goat antelopes group.

whitish underparts, a long bristly mane on its neck (ranging in color from white through brown to black), a short white beard and two short, sharp horns 6-10 in. long. Much larger than the goral, it measures about 60-70 in. head-to-tail and may weigh as much as 300 lbs. During the day it shelters among dense woodland vegetation or in a secluded clearing, emerging in the early morning and late evening to feed.

THE GOAT ANTELOPES — CAPRINS

ABOVE The distinctive hooked horns of an adult male chamois contrast with the undeveloped horns of the young animal beside it. The adult's horns are fearsome weapons in the fight for herd dominance during the rut. By getting its head beneath an opponent, the aggressor can rip upward into its belly, with fatal results.
ABOVE RIGHT A low-ranking chamois (right) avoids any damaging confrontation by adopting a cringing posture before a dominant individual.

Although the mainland serow has a clumsy-looking, rather slow gait, it is a skilled climber, able to run nimbly up almost vertical rocky slopes using the smallest irregularities in the rock surface as footholds. It is an aggressively territorial animal and will attack other serows that enter its feeding zone. Both sexes mark their home range with scent from their preorbital glands (in front of the eyes). It is also well able to fight off predators such as leopards, tigers, wild dogs and birds of prey that may attack the young. But it stays clear of man and can now be found only in impenetrable areas far from human settlement.

A target for hunters

The mainland serow has also been—and indeed still is—hunted mercilessly for its meat (although this is of poor quality), for its hide and even for its small horns, which are considered good luck charms. In Sumatra, where the local subspecies is now greatly reduced in number, it is legally protected, but its best protection is the inaccessibility of its mountain habitat.

The Japanese serow is a little smaller and lighter than its mainland relative, measuring 50-65 in. head-to-tail, some 24-36 in. at the withers and weighing 90-140 lbs. Its coat varies in color from dirty white to gray or brown, and in some regions completely black or dark brown individuals can be found. The animal is distributed throughout the Japanese islands of Honshu, Shikoku and Kyushu, but is now confined to a few restricted areas of mountainous country. It prefers steep areas covered with woodland or thick bushes, where it feeds on grass, shoots and leaves.

Japanese serows were quite abundant at one time, but the destruction of the forest and uncontrolled hunting have taken their toll (even though it was common for hunters' dogs to be gored to death by their quarry's sharp horns). Until recently, the meat and hide were highly prized, and the horns were even used for harpoon points by local fishermen. Fortunately, the Japanese government declared the serow a "national monument" in 1934, and in 1955 it was afforded complete protection. Despite this, poaching is still common, and the erosion of its forest habitat continues.

THE GOAT ANTELOPES—CAPRINS

SURE-FOOTED CLIMBERS

A plucky fighter

The Japanese serow's short horns have always proved an effective weapon—used against such enemies as the Japanese wolf (before it became extinct). Its horns also discourage other serows from eating its food resources; if it finds an intruder within its small feeding territory (usually 5 to 40 acres), it will immediately attack and chase it off. Serows are even capable of fighting off predators, including Asian black bears. The serow has well-developed scent glands in front of its eyes and between the toes, and it uses these to mark its territorial boundaries throughout the year. The foot glands scent-mark the serow's regular trails while it rubs the facial glands against twigs, plant stems and rocks. The serow also marks its territory with dung heaps.

The Taiwanese serow is a dwarf subspecies of the Japanese serow that, today, is restricted to the rockiest, most impenetrable parts of Taiwan. It is a lighter color than its Japanese relative and both sexes have large cheek beards. Its habits are similar to those of the Japanese serow—although it prefers more densely wooded areas. Split into a number of very localized populations, it is endangered by deforestation and uncontrolled hunting. Like all isolated island species, the Taiwanese serow is vulnerable to exploitation, and its future looks bleak.

A Rocky Mountain goat

The mountain goat of North America has a striking creamy-white coat. The coat has two layers: an outer layer of stiff hairs that forms a thick mane on its neck and on the upper parts of its legs; and an undercoat of long, thick wool to keep out the cold. A sturdily built animal, it is similar in size to the mainland serow, from

BELOW Outlined by the low light of an Alpine winter, a young male chamois peers over a cliff edge into a deep valley. Chamois can leap down almost vertical cliffs with confidence, and find footholds on apparently sheer rock faces and treacherous scree slopes, enabling them to escape less sure-footed predators such as bears and wolves.

THE APENNINE CHAMOIS
— NIMBLE CREATURE OF THE MOUNTAINS —

A rare subspecies of chamois is found in the Apennine mountains of central Italy. The Apennine chamois has long, sturdy horns, slender limbs, and a longer tail and less thick-set body than the other subspecies. Its winter coat is dark brown with broad, white or dingy yellow patches on the shoulders, hindquarters and throat, but its summer coat is very similar to that of the other subspecies. At one time this animal probably occurred throughout the Apennines, but today it survives only on a few mountains within the Abruzzo National Park, where some 400 individuals are found together with natural predators such as the wolf and the golden eagle.

Until recently the survival of the Apennine chamois seemed unlikely, as numbers were declining dramatically owing to poaching and the negligence of the Abruzzo park authorities. No more than 30 to 40 animals were left after World War I. A subsequent slight increase in numbers was followed by another ruinous decline during World War II, and the subspecies came perilously close to extinction.

Since the early 1970s the chamois flocks have increased, although they are still plagued by poachers. Other disruptive factors include competition with domestic livestock such as sheep and goats, the presence of sheepdogs (some of which have returned to the wild) and the tourist invasion during the summer. There is also the ever-present danger of an epidemic, which might easily wipe out the whole population. To prevent this from happening, the Apennine chamois needs to be reintroduced to parts of its former range where new breeding populations would preserve the subspecies in the event of a disaster in the Abruzzo.

The Abruzzo chamois spends the cold months from late autumn to the end of spring sheltering in woodland at lower altitudes. As spring turns to summer, they climb up to the alpine meadows and stay there until the first signs of winter appear.

In April or May, females without offspring join up with the animals born the year before and head for the alpine meadows, where there is plenty of food; pregnant females wander off alone to the steep, rocky parts of the mountains, where they give birth in late spring. Mature females normally bear one kid each year—although twins and even triplets occasionally occur.

Evading eagles

Mothers spend most of the time close to their young on the rocky crags where they can find protection from eagles and wolves. Sometimes they will join up with groups of young from the previous year and the older, sterile females on the meadows. This happens more often as the young chamois get stronger and more independent. Adult males spend the summer in the woods on their own or in small groups, though young males often mingle with the females.

At the beginning of autumn the herd size starts to increase markedly and, at the end of September, the adult males start to leave the woods and enter the meadows, surrounding the areas containing the females and their young offspring.

Courtship does not start until the beginning of November, when the mature males display the blackish-brown and white winter coat characteristic of the Abruzzo and Pyrenean chamois. As the coat color of the adult males changes earlier than that of immature animals, breeding males can be spotted easily, even from a distance. They often show off their powerful profiles to potential rivals by climbing onto higher rocks.

Fully mature males are dominant over all age groups and become increasingly intolerant of young males, chasing them out of female herds as the height of the rut approaches. The resident male also has to prevent females from entering the territory of other males. He will threaten an escaping female with lowered horns in order to force her back to the center of his own territory. Rounding up the females is an exhausting test of strength, and the males are not always successful in keeping all the females for themselves.

When the rut reaches its peak in mid-November, the males approach the females with their necks and muzzles held aloft to show off their white throat patches and to hide their threatening horns. If the female is in heat, she leaves the herd and is chased among the crags and gullies until the pursuing male finally succeeds in mating with her—provided he is not interrupted by any rival males.

FAR LEFT A small group of Apennine chamois cluster on a rocky ledge high up the mountainside. Chamois are sociable animals, and the females spend much of the year foraging in flocks of up to 30 animals.
BELOW LEFT A solitary male chamois searches for food beneath the snow cover. Adult males usually live alone until the breeding season, when they join the females and gather harems.
BELOW During the rut the males are highly territorial, marking bushes with their facial scent glands (top) and chasing off potential rivals (bottom).
BELOW RIGHT Chamois are able to climb safely onto the most perilous of cliff edges as their hooves have a thin, hard outer rim that finds a hold on rocks and cracks, while the sole is covered by an elastic pad that stops the animal from slipping.

THE GOAT ANTELOPES—CAPRINS

ABOVE A small flock of chamois file across a snowfield well above the tree line in their search for high mountain pasture. Young chamois rarely stay long in one area, but move from place to place to take advantage of seasonal food sources. Older animals are more settled, and in winter they prefer to feed at lower altitudes where the grazing is better and the climate not so harsh. Chamois have been known to survive after fasting for up to two weeks when the snow lay so deep that they could not find any food.

45-65 in. long, weighing some 310 lbs. at most and measuring up to 4 ft. at the withers. The mountain goat's sharp, black horns are longer than the serow's, measuring up to 12 in.

The mountain goat lives in the alpine and tundra zones high on the mountain slopes of the Rockies and other ranges, from southeastern Alaska through the Yukon Territory and down to north-central Oregon, Idaho and Montana. Although hunting has reduced populations in some areas, the mountain goat has been introduced to many places, including the Alaskan islands, Colorado and the Black Hills of Dakota. The total population is probably about 100,000. The cold, inhospitable habitat of the mountain goat includes steep cliffs and the edges of great glaciers, and is often blanketed in snow.

Mountain goats feed in small groups of up to four or five animals for much of the year, and in large gatherings in winter when food is scarce. While less jealous of its feeding territory than the serow or goral, aggressive encounters between the members of a group are frequent. According to one study, each individual is involved in an average of six to seven disputes (of varying intensity) per hour. It seems that among mixed-sex groups, in some populations at least, adult females with young dominate the males.

The mountain goat feeds in the early morning and late afternoon, grazing on woody plants, herbs and grasses during the spring and summer months. In the winter it eats conifer shoots, mosses and lichens.

Gripping feats

A skilled climber, the mountain goat is capable of impressive acrobatic feats when clambering among the rocks, crevasses and scree slopes of its mountain habitat. It owes its sure-footedness to a number of special anatomical adaptations. The two parts of each hoof can be closed up or splayed out at right angles, depending on the terrain; on narrow ledges, for example, a closed hoof gives a better grip, but in deep snow it helps to have broader hooves to spread the weight over a wider area. Its foot pads, moreover, are

THE GOAT ANTELOPES — CAPRINS

RIGHT With its clumsy build, broad head and lateral horns, the takin is superficially more like a cow than a goat. Although it has no specialized scent glands, it leaves its scent wherever it goes, owing to an oily, acrid secretion that permeates its coat.
PAGES 546-547 Barbary sheep are large-horned inhabitants of the rocky mountains and desert hills of North Africa. The males are distinguished by the luxuriant mane that hangs from the throat and chest down to the front legs as far as the knees. Barbary sheep take their name from the western region of North Africa once known as Barbary, after the Berber (in Arabic, *barbar*) people who lived there.

elastic and horny—not hard as in deer—and this gives a better grip on bare rock surfaces and ice. Two rudimentary toes flanking each cloven hoof prevent the goat from slipping when it moves downhill.

Mountain goats can live to an age of up to 18 years in the wild. They have few enemies except for man, who has hunted the goat in the past and continues to hunt it—often shooting the animal to no purpose, since the injured or dead goats frequently fall from the rock face into gullies or crevasses that the hunter cannot reach. Wolves hunt them, but the goats do not make easy prey; a goat pursued by wolves will seek immediate refuge among sheer rocky crags where a wolf cannot follow. The goats also defend themselves against all manner of predators—including eagles—by using their sharp horns. Only during winter, when food is short and the goats are weakened, do predators stand a good chance of capturing an adult. The young, however, are vulnerable to various predators, such as pumas, brown bears, wolves, coyotes and eagles.

Resistant females

Mating activity starts in October, reaches a peak in November and then declines, ceasing in early January. Mixed groups of adult males and females form at the start of the mating season, the males becoming more than usually intolerant toward other males. Their first attempts at courtship are hesitant. At first the females react to the advances of the males with threats and blows with their horns, since they do not become receptive until some time after the mating season begins, and the males often have to beat a retreat.

As part of their courtship ritual, the adult males dig out holes in the ground by striking it with their front hooves. Since they generally urinate in the holes and lie down in them, their white coats soon become filthy and begin to give off a powerful smell. The younger males are not as wholehearted about this activity as their seniors, and their coats remain much cleaner. However, it is the dirtiest and smelliest animals that achieve most success with the females, suggesting that this behavior shows females, and other males, which individuals are dominant and in a condition to breed.

Seduced by smell

When the females are approached by the pungently scented breeding males, they are much more subdued than they are at the beginning of the breeding season. The male checks each female by smelling her genital area to see whether she is in heat.

If two adult males find themselves in competition, they indulge in an impressive threat ritual. They position themselves alongside one another, with their heads facing in different directions and held so low as to be almost hidden between their forelegs. They then raise their shoulders to appear more frightening. Occasionally, neither animal gives in at this display of physical strength, and the encounter ends in a fight— an exchange of horn blows aimed particularly at the underside and hindquarters. Mountain goats have thick skin on their rumps for protection, but severe, even fatal, injuries may still occur during fights.

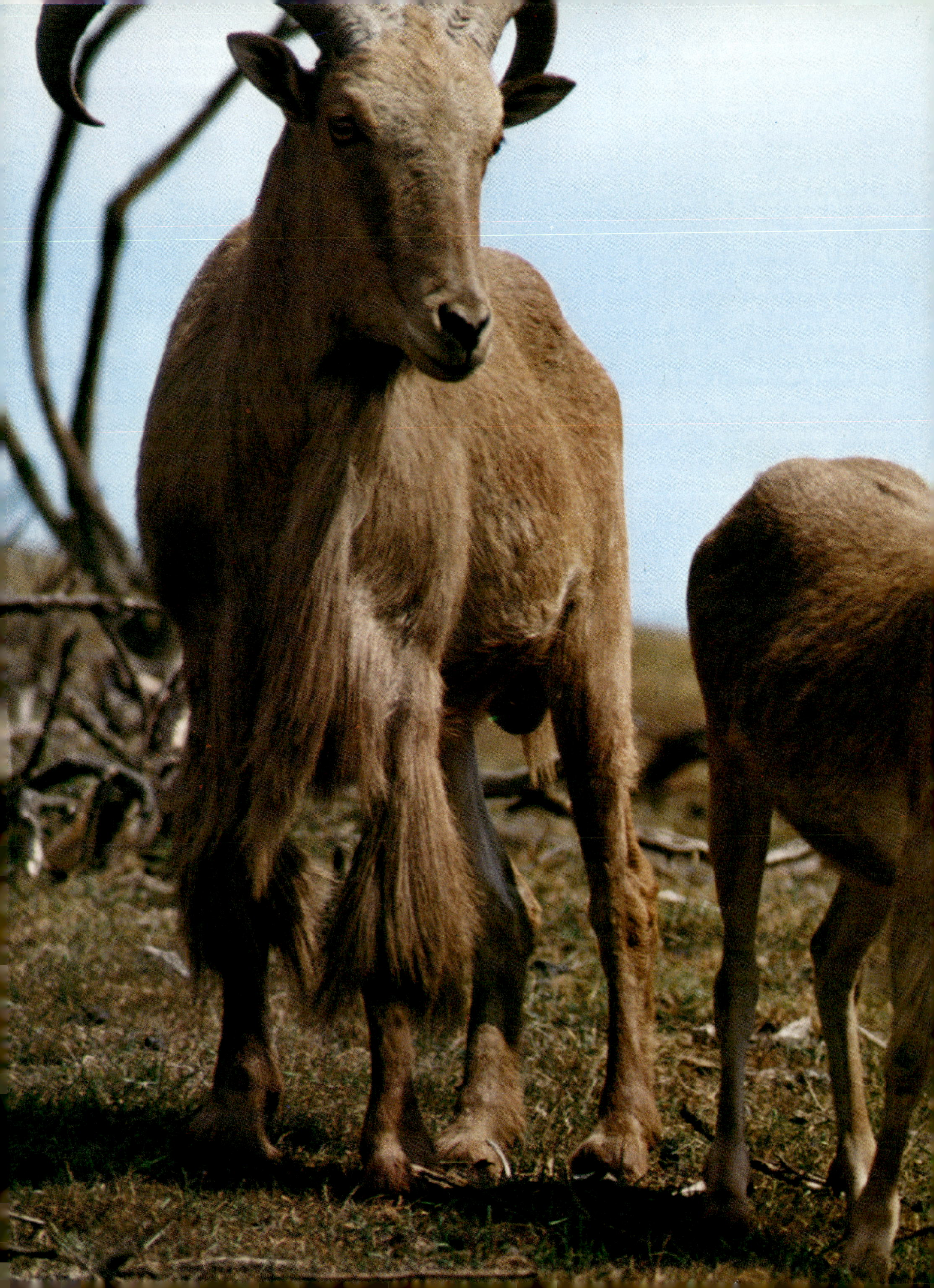

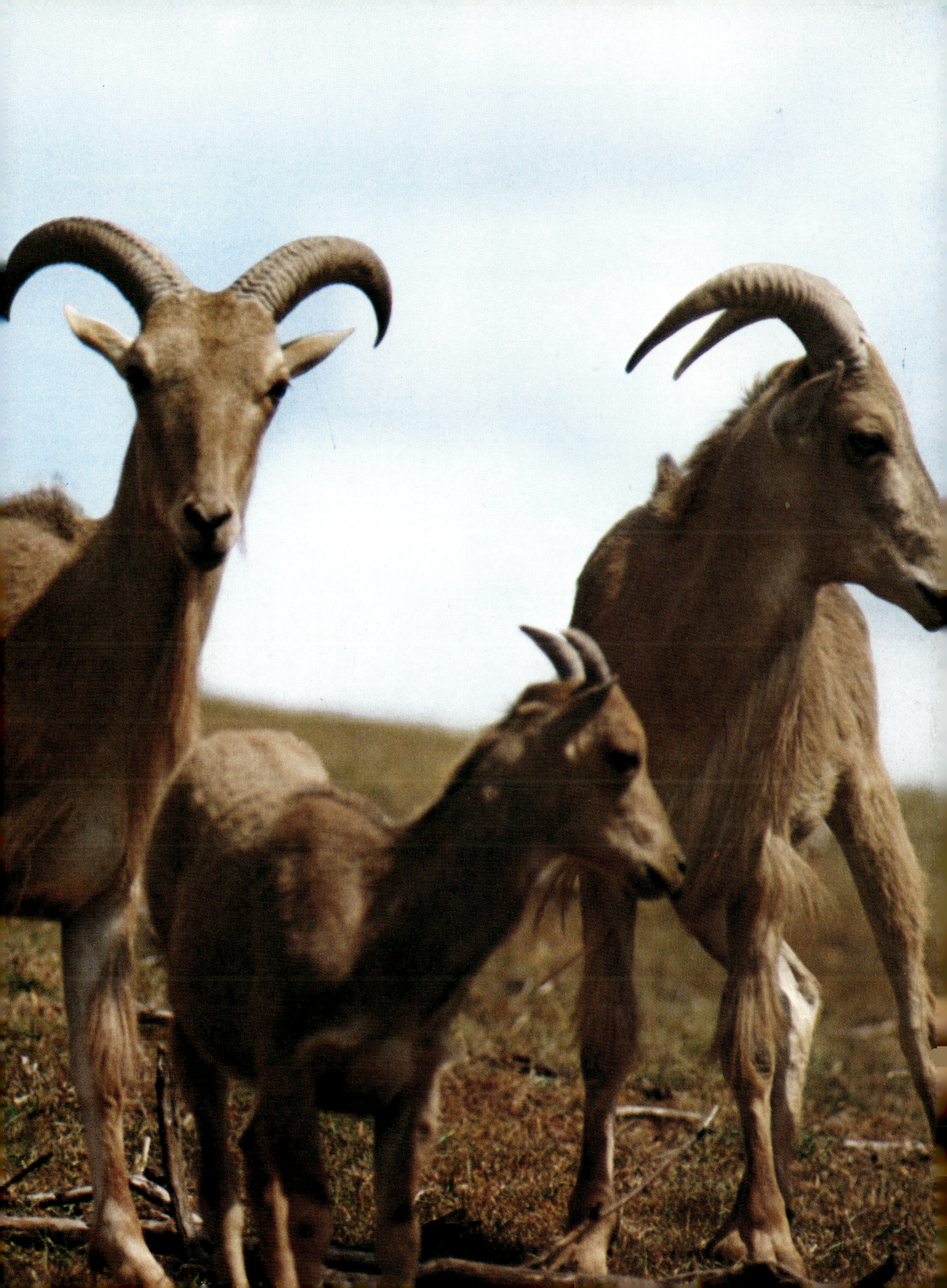

THE GOAT ANTELOPES—CAPRINS

LEFT Shaggy hair matted with ice, a female musk ox forages for food with her calf in the Arctic tundra. As vegetarians, these massive animals cannot survive in areas of permanent snow cover. They feed on the lichens, mosses and small plants that grow at the edge of the glaciers and ice sheets.

An agile mountaineer

The chamois is a small, agile inhabitant of high mountains, extensively hunted for its skin (used to make "shammy" leather), meat, and fur (used for the tufts on traditional Tyrolean hats). In winter, it has a black or dark gray coat, but in summer this changes to tawny-brown with a black stripe along the back. Its face is white with two black stripes running from the base of each horn to the muzzle—rather like the mask of a badger—and its short horns, up to 8 in. in length, are sharp, slender and swept back at the ends to form two little hooks. The chamois grows to 4 ft. long with a tail up to 16 in. long and height at the withers of about 30 in. It may weigh as much as 110 lbs.

There are nine subspecies distributed throughout the mountain chains of Europe and the Near East (the Cantabrians, Pyrenees, Alps, Apennines, the Tatras of Poland and Czechoslovakia, Carpathians, the Balkan mountains in Bulgaria, the Anatolian mountains of northeast Turkey, and the Caucasus).

Chamois are renowned for their sure-footedness and agility as they leap across rock faces and boulder-strewn mountainsides. They are usually found on steep mountain slopes broken by clearings and extensive mountain meadows, although they will also visit sheer, craggy cliffs covered with dwarf conifers. Females with young tend to stay within the shelter of woodland during the cold winter months. Chamois are most active during the day, but they also come out to feed at night, particularly in summer.

Whistles and stamps

Chamois are gregarious animals. Females and young commonly live in herds of 15-30 individuals, led by senior females, particularly during summer and autumn. One of the females acts as a sentinel, warning the feeding herd of danger by stamping her feet on the ground and uttering a sharp, high-pitched whistling sound. In winter, when food is thinner on the ground, herds tend to split up as each animal searches for foliage, lichens, mosses and patches of grass amid the snowbound forests and mountain slopes. The herds re-form in late spring.

Although younger males often join the herds of females and young, fully adult males are solitary and generally live in wooded, rocky country for much of the year. In summer, they will sometimes defend feeding territories as serows do, safeguarding their food supplies in preparation for the exertions of the mating season. In autumn, they move out of the woods to join the females.

Adult males are normally fairly tolerant of one another during the summer (unless they are defending their food resources) but during the autumn rut they become deadly rivals. The chamois mating season is marked by pursuits of the females and threats against rival males as the males attempt to gather and defend harems of females.

Might is right

The outcome of a confrontation between rival male chamois usually depends on rank, and this in turn is determined by body bulk. Animals of equal rank will try to intimidate one another by taking up a curious arched-back posture, with the hair on their spines standing up and their heads stretched upward. If this fails to impress, each animal points his horns toward his rival, and may even charge him and try to hook his

548

THE GOAT ANTELOPES—CAPRINS

horns into his throat or underside. Chamois are quite capable of ripping their opponents open, and real fights often prove fatal. Even if one runs away, he may be ruthlessly pursued.

Usually, the two contenders avoid killing each other by restricting themselves to a series of threats of varying intensity. If an attack seems inevitable, it is sometimes warded off by a display of submission: one of the rivals crouches down with his neck and head parallel to the ground, and attempts to appear as humble as possible. He may even lie down, which makes it impossible for the aggressor to attack his underside with his hooked horns.

Individuals of all ages scent-mark bushes, stems and twigs within their home ranges, using a secretion from the occipital scent glands (on the backs of their heads). Mating males are the most enthusiastic scent-markers, rubbing their glands vigorously against any surface they encounter. The scent marks serve not only to mark out territory, but the act of marking is an indirect threat when carried out in the presence of others. In some cases, chamois actually use their horns to attack the bushes and shrubs they are scent-marking, especially when the animals are sexually aroused or frustrated. They are quite long-lived animals, and may survive for up to 20 years or more in the wild.

ABOVE **When threatened by predators, musk oxen stand their ground and form a defensive wall, with the adult males at the front and the females and young safely behind. The males' sharp horns present a formidable barrier, and the animals may charge and trample predators that draw in close. Such tactics usually work well, even against a pack of wolves, but they are disastrous against humans armed with guns, for whom the stationary groups of musk oxen make easy targets.**

The isard or Pyrenean chamois is still abundant in some parts of the Pyrenees and is similar in all respects to the Apennine chamois except that it is much smaller. The subspecies found in the Cantabrian mountains of northern Spain is smaller still. It also has short, slender horns that curve first forward and then backward, ending in sharp points that diverge widely.

Asiatic takins

The takin and musk ox are massive, clumsy-looking creatures that are halfway between goats and cattle in appearance. The more goat-like of the two species is the takin, an inhabitant of the temperate mountain bamboo forests of central China, southeastern Tibet, Bhutan and northern Burma, at altitudes of between 8200 and 13,000 ft. They weigh up to 770 lbs., measure

THE GOAT ANTELOPES—CAPRINS

GOAT ANTELOPES CLASSIFICATION: 2

Rupicaprini

The Rupicaprini was the earliest of the Caprinae tribes to evolve. Members of the tribe have short, sharp horns that they use for defending feeding territories. They range over most of the mountainous regions of Eurasia and North America.

The genus *Capricornis* contains two species: the mainland serow, *C. sumatrensis*, which inhabits forested slopes in eastern and Southeast Asia; and the Japanese, or Taiwanese, serow, *C. crispus*, of Japan and Taiwan. The gorals inhabit steep, rocky slopes with varying degrees of vegetation cover, over much of eastern Asia south to Thailand and west across the Himalayas. Evidence now suggests that there are two species: the common goral, *Nemorhaedus goral,* and the red goral, *N. cranbrooki.* The mountain goat, *Oreamnos americanus*, lives high in the Rockies and other mountain ranges in Canada and the USA, while the chamois, *Rupicapra rupicapra*, is native to the mountains of central and southern Europe and southwest Asia.

Ovibonini

The tribe Ovibonini comprises just two species, the takin, *Budorcas taxicolor*, and the musk ox, *Ovibos moschatus*. They are bulky animals with long hair and short, stout legs. Takins inhabit upland bamboo forests in China and the eastern Himalayas. Musk oxen are animals of the far north, ranging over the Arctic tundra from Alaska to Greenland.

---

about 4 ft. long with a 4-in. tail and stand 40-50 in. at the withers. Takins have shaggy, oily coats that vary in color from light fawn to dark red and blackish-brown; one race has a golden-yellow coat, giving rise to the name "golden-fleeced cow." The animals have short, thick hind legs and longer, sturdy forelegs, massive muzzles with dark noses and stubby backswept horns.

Takins are shy, retiring creatures that spend most of the day hidden among trees and shrubs. Emerging at dusk to graze, they may feed all night before seeking cover again at the first sign of daybreak. If they are startled by intruders or predators they make a quick dash for the safety of the dense forest. During the hot season, the females and immature males live in large groups at the upper edges of the forests or in the high alpine meadows, but during the winter the big herds break up into smaller groups that descend to grassy valleys at lower altitudes.

Adult males are usually solitary, and rarely seek the company of other takins except during July and August, over the breeding season. They mark their territory by scent, but instead of producing smelly secretions from specialized skin glands, they impregnate their surroundings with the strong smell of their oily coats. The young are born in March to April, females often entering the thick forest to give birth. A single young is usual, and twins are rare. The young takins are well-developed at birth, and able to follow their mothers at three days old.

Takins of Tibet and China

Of the various subspecies, the male Tibetan takin has a fine golden-yellow fleece during the summer that changes to iron-gray in winter. The female is much grayer in color. The Tibetan takin inhabits the Szechwan region of China, and was very common until the first part of the 20th century. A dramatic fall-off in numbers due to hunting prompted the Chinese authorities to place it under strict protection, and numbers are now recovering.

Another subspecies, the rare Bedford's takin, is found in the southern part of the Shensi region of central China. The male has a dark golden coat, while that of the female is more creamy in color. It lives amid craggy rocks and steep slopes below the tree line, favoring dense shrub thickets, and moves over the ground very nimbly in spite of its bulky appearance. Today the entire wild population of Bedford's takin amounts to only a few hundred closely guarded individuals, and takins are still hunted for their meat over much of their range.

THE GOAT ANTELOPES—CAPRINS

ABOVE **The Himalayan tahr prefers rocky terrain and establishes its territory in the roughest, most inaccessible mountain areas. While the Arabian tahr and the Nilgiri tahr are low in number with localized ranges, the Himalayan tahr is widespread across the Himalayan range. It has been successfully introduced to New Zealand where it is thriving in the absence of predators or natural competitors.**

Giant of the tundra

The musk ox is a huge, robust animal that can withstand the intense cold of the bleak Arctic tundra. Standing up to 5 ft. high at the withers and weighing as much as 900 lbs., it has a long, shaggy coat consisting of two layers of hair. The outer layer is made up of coarse, dark brown guard-hairs, reaching almost to the ground. These protect the inner coat of fine, soft, pale brown hairs, which is so dense it is able to keep out the bitter cold. The animal's most striking features, apart from its sheer bulk, are its massive head —armed with sharp, curved horns growing from a flattened base between the ears—and the noticeable hump at the shoulders. The animal gets its name from the strong smell given off by the facial musk glands of the male in rut. Females are smaller, on average, than males and have smaller horns.

Musk oxen live in herds of varying sizes: these usually comprise 15-20 animals in winter and about 10 in summer. Occasionally, up to 100 animals may gather together. When a herd is alarmed by the approach of a wolf or a human intruder, the animals quickly form a circle, with the adult males on the outside facing the danger. However, the herd may try to beat a hasty retreat to a safer position, such as a steep slope, before adopting this defensive formation.

When forming the circle, the females and younger animals retreat to the center, the youngest and weakest animals often sheltering between the legs of the females. The herd leader is almost always an adult male, who controls the defensive action and keeps the circle intact. Every now and then, one musk ox may make an intimidating dash at the predator, attempting to make an upward strike with its sharp horns.

These defensive tactics work well against a wolf, but when the predator is a human with a firearm, practically the worst thing the animal can do is to retreat to an elevated position and form a circle. Since musk oxen allow hunters to approach to within a few yards once they have formed their defensive circle, it is not difficult to destroy an entire group of the animals. Inevitably such tactics led to a serious decline in numbers, even when primitive firearms were used.

THE GOAT ANTELOPES—CAPRINS

ABOVE Perched on a rock face high above the tree line, a male Alpine ibex demonstrates its extraordinary head for heights. Not only can ibex cope with the most treacherous of mountain environments, they can also withstand the depths of winter at high altitudes, often moving on to steeper slopes where less snow accumulates. They seek cover from blizzards behind rocks and under snow ledges, and only retreat to the shelter of valleys during very harsh weather conditions.

New lease of life

Musk oxen once ranged from Alaska to Greenland. They (or a very closely related species) also occurred in the Arctic regions of the Old World until about 2000 years ago. But musk oxen were exterminated in Alaska and nearly wiped out on the Canadian mainland during the 19th and early 20th centuries as the Arctic regions were opened up by explorers and hunters. By the 1930s, only about 500 musk oxen survived on the Canadian mainland. Conservation programs have now allowed the world population to build up to an estimated 25,000 animals. Small populations of musk oxen have been successfully reintroduced to Alaska, northern Quebec, the USSR, the island of Spitsbergen (in the Arctic Ocean), and Norway.

Lone individuals (usually old males) spend much of their time in rocky areas. Though they may look clumsy, they can run away from danger with surprising swiftness and agility even in snow. If cornered, however, they stand with their backs to the rock to face their aggressor.

The mating season lasts from July to September, with a peak in August. Although musk oxen begin to nibble vegetation when they are only a week old, the young may not be fully weaned until they are a year old. One-year-old animals remain with a group of females and their new offspring for at least another full year—young sub-adult males are then expelled from the herd by the oldest male, who controls the herd. They cannot rejoin the herd until the mating season has finished, instead forming all-male groups that roam and feed independently of the mixed herd.

Duels to the death

Fights between rival males over a female are spectacular and often end in the death of one of the contenders—up to 10 percent of adult males are killed in such duels at the height of the mating season. The two males threaten one another with loud bellows, at the same time showing off their bulk. If neither animal is put off by this display, they charge with lowered heads, butt one another and try to pierce the opponent's throat with their sharp horns.

The high-living tahrs

Tahrs are medium-sized, short-tailed goat antelopes living in mountainous country. Both sexes have horns, but those of the male are both longer and heavier than

THE GOAT ANTELOPES—CAPRINS

ABOVE The ibex is one of the easiest of the goat antelopes to recognize, owing to its magnificent scimitar-shaped horns. In some subspecies the males may be crowned with horns over three feet in length.

LEFT A male courts a female with his neck and muzzle outstretched and held parallel to the ground (A). During clashes between rival males, each animal tries to gain an advantage by shifting on to higher ground (B).

THE IBEX

Ibex have declined in numbers over most of their range, largely because they have been subject to a long history of indiscriminate hunting at the hands of man. In the second half of the 19th century the ibex almost disappeared from the entire Alpine mountain chain in Europe. Fortunately, the animals found a refuge in the Gran Paradiso Reserve in Italy. The reserve has since become a stronghold for the Alpine ibex, and numbers there now stand at about 5000. Using this stock, efforts have been made to repopulate many other parts of the Alps and the mountains of Yugoslavia. With careful protection the numbers of ibex in these areas are now increasing, but continued protection is essential if the populations are no longer to be regarded as under threat.

Another seriously endangered subspecies is the Walia ibex, which lives only in the Simien Mountains in Ethiopia. In the mid-1970s the total population stood at about 300. Of this total, some 240 were living in the Simien Mountains National Park, where, in theory, they are protected by law—in practice, poaching continues.

A closely related species, the Spanish goat or Spanish ibex, has also been severely affected by overhunting. The subspecies, which lives in the Pyrenees, is currently on the very edge of extinction. No more than 20 individuals are thought to survive—when animal populations fall to such low numbers, they are highly susceptible to natural threats such as disease or climatic hazards, let alone the threat brought by human actions.

THE GOAT ANTELOPES—CAPRINS

ABOVE Ibex are among the most nimble of all the mountain-dwelling goat antelopes, able to run along narrow ledges and jump across ravines several yards wide. Their hooves are adapted for gripping steep and slippery rock surfaces, but ibex avoid glaciers and snowfields, where even they cannot find an adequate foothold.

the female's. The female is substantially smaller than the male—with about 60 percent of the male's body weight. The three isolated populations of tahr—the Himalayan, the Nilgiri and the Arabian tahr—form three distinct species.

The Himalayan tahr is the biggest species—a male may weigh as much as 220 lbs. Standing 20-40 in. tall at the withers, it measures 4-6 ft. from head to tail. Himalayan tahr live at altitudes above 10,000 ft., but below the tree line, in the steep foothills of the Himalayas in northern India, Nepal and the remote Indian state of Sikkim. They are rare in India, being hunted both for their skin and meat, but they have been introduced to South Africa and New Zealand, where they have adapted well to their new homes.

The Himalayan tahr has a long coat, especially in the male, which also boasts a shaggy mane around its neck and shoulders, extending to the knees. Its color varies from reddish-brown to almost black.

The tahr's mating season is in winter, from late October to January, when males fight one another for females. They assert dominance over rivals by pointing their noses in the air and stretching out their necks to erect the thick, lighter-colored mane. This makes the animal look larger and may be enough to scare off the rival. If it does not succeed, the animals may end up head-butting, but this is half-hearted compared to the blows exchanged by other goat antelopes.

Nilgiri tahr

The Nilgiri tahr takes its name from the Nilgiri hills in southwestern India where it lives. It is also found in a national park in Kerala state in south India. Although the Nilgiri tahr is now strictly protected, its numbers have dwindled to less than 2000. It prefers to live at altitudes of 4000-6000 ft., especially in areas where there are steep cliffs. Pressure from a fast-increasing human population and the development of the Nilgiri tea plantations have not helped this animal to increase in number.

Shorter than the Himalayan tahr, the Nilgiri tahr measures 20-40 in. from head to tail and weighs up to 220 lbs. It is short-haired, the males being almost black with a silver saddle and females being yellowish or grayish-brown with white bellies.

THE GOAT ANTELOPES—CAPRINS

Mating may occur at any time of the year, but it usually takes place in summer, with births following in winter after a six- to eight-month gestation period. Nilgiri tahr are gregarious animals, forming herds of 50-100 individuals. Adult males sometimes cut themselves off from the rest of the herd, but they are not generally aggressive toward other mature males. Young males, however, are more irritable and often exchange head-butts. The adults confine their fighting to the mating season, pawing the ground with their forefeet, then rearing up on their hind legs to balance for a second before directing all their falling weight into a head-butt.

The smallest of the three species, the Arabian tahr is classified as an endangered species by the IUCN (International Union for the Conservation of Nature and Natural Resources). It lives only on rocky precipices in the mountains of Oman in the Middle East. Standing about 2 ft. tall at the shoulders and measuring less than 3 ft. from head to tail, it weighs only 65 lbs. at most. Males lack manes and have short, flattened horns about 10 in. in length. They have a dark stripe along their backs and on their muzzles. Their grayish to tawny-brown coats consist of brittle, short hairs that form two large tufts at the base of the muzzle and a fringe over the hooves.

ABOVE With its long corkscrew horns, the markhor is one of the most distinctive goat antelopes of Asia. An inhabitant of remote mountain areas, it was unknown to zoologists until the 19th century. Its name appears to be derived from the Persian for "snake-eater," but evidence of any such diet has yet to be discovered.

The high-altitude ibex

The ibex is a large goat and one of the best-known of all mountain animals. It is a strong, sure-footed inhabitant of mountain ranges across the breadth of Eurasia, from the European Alps through Syria, Arabia and northern Ethiopia to Afghanistan, Kashmir, Siberia, Mongolia and central China. Its most striking feature is its pair of thick, backward-curving horns, with prominent knots on the front surface, that can grow to as long as 50 in. in the male (although they are much smaller in the female).

There are several subspecies of the ibex, including the Alpine ibex; the Nubian ibex of North Africa, the Near East and Arabia; and the Walia ibex (now seriously endangered), which lives only in the Simien Mountains of Ethiopia.

Ibex usually confine themselves to high mountain regions above the tree line, and are able to move around on the steepest slopes, even across the face of

THE GOAT ANTELOPES—CAPRINS

The goat antelopes range over most of Asia and parts of Europe, North Africa and North America.

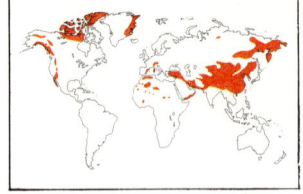

sheer cliffs. To improve their grip on steep surfaces, the front of their hooves are equipped with sharp points, which can be dug into the rock face, rather like mountaineers' crampons. In the vast Himalayan range, herds of ibex may live at altitudes of 21,000 ft. or more, grazing on grasses and shrubs, tree bark, roots, mosses and lichens.

Ibex weigh up to 330 lbs., reach 70 in. in length and may stand up to 3 ft. tall at the shoulders. They are gregarious animals, but males and females spend most of their lives apart, forming separate herds of males or females and young. These herds mingle only during the breeding season. Female herds usually number up to 40 individuals, led by an old, dominant animal. Male herds may be larger, and the dominant animal is the strongest rather than the oldest member of the group. When old males become too weak to fight for females in the breeding season, they leave the herd to lead a solitary life.

Crashing horns

In the Alps, the rut takes place in December and January. During this period the sexes mix freely, and males wander among the female herds in search of partners. Rival males confront each other by rearing up on their hind legs and exchanging violent butts with their horns until one recognizes the other's superior strength, and backs off. The rattle of horn crashing against horn echoes around the surrounding mountains, but despite the startling sounds and the dramatic spectacle of such battles, the animals are rarely hurt. The only protracted fights occur when the supremacy of the dominant male is challenged by another male of similar strength and weight.

Dangers of the thaw

Apart from man, the main threat to the ibex comes from rockfalls and avalanches, especially in the spring when the snow and ice start to thaw. Adults may be killed by leopards, bears and wolves, and very young ibex also make easy prey for eagles, foxes and jackals.

The Spanish goat or Spanish ibex is a species of goat antelope found in the Pyrenees and the Cantabrian Mountains of northern Spain. It is similar in appearance to the Alpine ibex but slightly smaller, and its horns are different in shape. The horns are triangular in cross-section and, rather than simply arching backward like those of the ibex, they twist

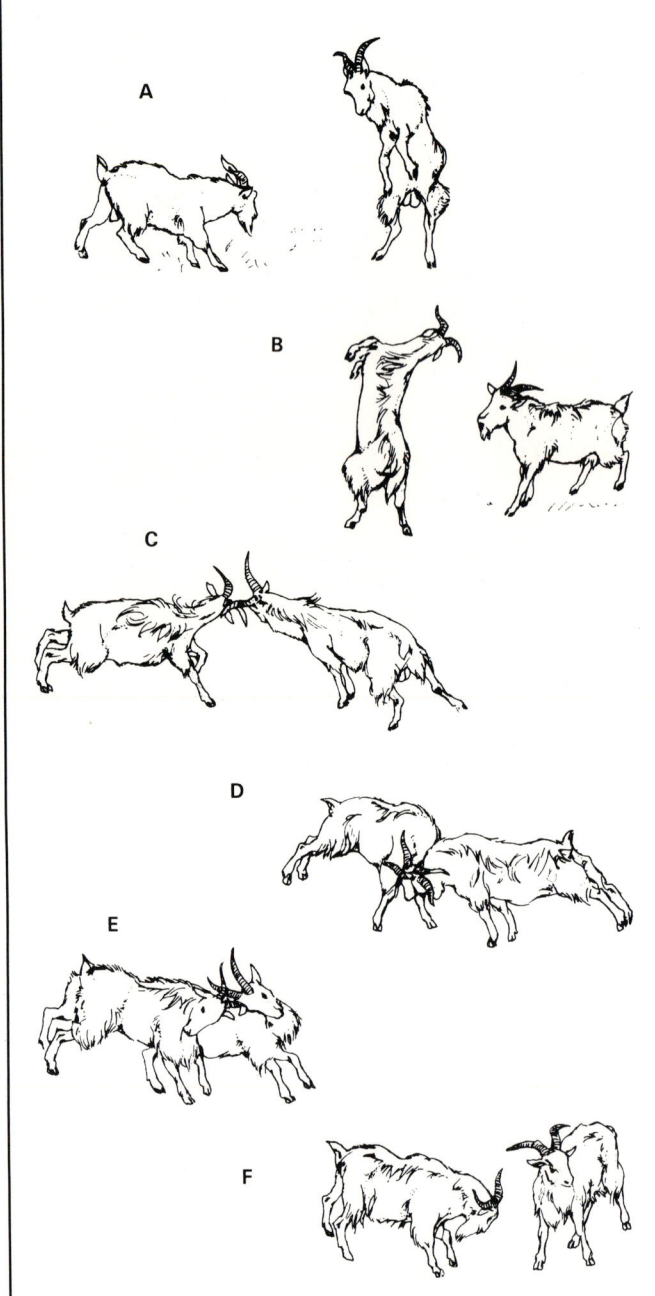

ABOVE A fight between wild goats: one male rears up on its hind legs in the attack position (A), then stands on all fours again while its opponent takes up the attack position in turn (B). The two animals then lock horns, grappling with each other, trying to twist the other's neck (C), (D) and (E). Finally, one animal breaks free, turns his head away in submission and gives up the fight (F). The clash of horns during a fight can be heard up to several miles away, echoing through the valleys.

THE GOAT ANTELOPES—CAPRINS

ABOVE Older male Barbary sheep fight for control over groups of females during the mating season in November. Apart from head-butting, males interlock horns and try to throw their opponent to the ground (top). They may also deliver sharp blows with their horns to the opponent's hindquarters (bottom).
RIGHT The Barbary sheep is the only sheep native to Africa and is hunted by the people of the Sahara for its meat, hide, hair and tough sinews (which they use as rope).

outward and upward, then backward and finally upward again, with an inward turn at the tip. Many local populations of the Spanish goat were much reduced or even exterminated by hunting in the 19th century, but there have been a number of re-introductions which have allowed some populations to recover their numbers.

Markhor corkscrews

The most distinctive feature of the markhor, sometimes known as Falconer's goat, is its long, tightly spiraled horns that can grow to well over 3 ft. in length. Males weigh up to 240 lbs. and stand up to 3 ft. tall at the shoulders. They sport a long, flowing beard, thick dorsal mane and long tufts of hair on their forequarters and thighs. Females are much smaller—reaching less than half the weight of males.

Markhor live in northern Pakistan, Afghanistan, Kashmir, Ladakh (in India), and in the mountains of the USSR bordering these areas. They inhabit arid mountain slopes at altitudes of 2000-11,500 ft.; at these elevations they avoid competition with the ibex, which live higher up in the same mountain ranges. Unfortunately, the markhors' magnificent horns have made the animals prized hunting trophies. Overhunting and loss of habitat (from deforestation and the creation of grazing land for domestic livestock) have combined to make the markhor one of the rarest of the goat antelope species—many populations are highly vulnerable; some are already extinct.

Markhor browse on a variety of green plants and shrubs, and when food is short they have been known to climb 25 ft. up into trees to feed on leaves. During the summer the animals need to take shelter

THE GOAT ANTELOPES—CAPRINS

ABOVE Adult male Barbary sheep are agile climbers, living in the more inaccessible rocky terrain of the North African mountains. Their thick coat protects them from the desert cold at night.

from the sun during the hottest part of the day. The chilling conditions of winter, on the other hand, force them to descend to lower mountain slopes—a habit which makes them especially vulnerable to shooting by upland villagers.

Mating takes place between December and January—adult males do not appear to be territorial and are content simply to join up with groups of females. However, once they have joined a group they will try to drive away any other males that approach. The most common threat gesture is a rapid upward thrusting of the horns; to make a more aggressive signal, the male directs his horns straight at his foe and charges. If a fight develops, the males rear up and clash their horns like battling ibex.

Once a female goes into heat, the male keeps close to her until she allows him to mate. He stands behind her with his muzzle outstretched and tail raised, or stands beside her with his muzzle raised to one side. Sometimes he lowers his head until neck and muzzle are parallel to the ground. If the female moves, the male usually follows her, stretching his muzzle and sometimes sticking his tongue out; he may even give her a gentle kick. Six months after successful mating, the females leave their groups to give birth alone.

Markhor have a high birthrate, with numerous twin births, particularly from females over eight years old. Since the markhor now survives only in small, isolated populations, there is a serious risk of inbreeding—unfavorable characteristics may become concentrated in the population. Inbreeding puts them at risk to the spread of hereditary weakness and disease.

Forerunner of the domestic goat

The wild goat (or bezoar) is the probable ancestor of the domestic farmyard goat. Truly purebred wild goats may no longer survive, but wild animals—close in appearance to the original species—still live in some parts of Afghanistan, Pakistan, Turkmenia, the Caucasus, Turkey, Iran and Iraq, and on some Greek islands. In many areas they have interbred widely with domestic goats or goats that have reverted to the wild.

Female wild goats that are more or less purebred have a light brown coat, white belly, a grayish muzzle, a black tail, a dark patch on the chest, and slender, curved horns that grow 8-12 in. in length. Adult males are much larger and may weigh over 200 lbs.—twice the weight of the female. Their muzzles and the fronts of their necks are dark and their backs grayish-white. A dark mark crosses their shoulders, a dark band runs across their flanks, and another dark band runs along their back. The horns of mature males are scimitar-shaped, and much longer than those of the females, sometimes curving 50 in. in length. The coats of old animals turn silvery-gray in color.

Herds of wild goats usually number no more than 20 animals, a strict hierarchy being maintained in the group mainly through displays of horn size. The horns act as a measure of the animal's physical strength, bypassing the need for direct physical contests—fights between males are rare.

Pungent aroma

Adult males show one characteristic typical of many goat antelopes—the tendency to urinate on themselves. The pungent scent of the urine on their coat acts as a clear warning to rival males and may even serve to attract females.

It is never easy to distinguish the forerunners of domestic animals, partly because they are often

THE GOAT ANTELOPES—CAPRINS

ABOVE A young Barbary sheep nestles close to its mother. The offspring weighs from as little as 5 oz. to 7 lbs. at birth and suckles for three to four months.
RIGHT The grooming behavior of a Barbary sheep is similar to that of a domestic sheep: it uses its hind legs to scratch its face and neck (A and B); rubs its long neck mane on the ground when it is itchy (C); uses its horn tips to scratch its back (D); and licks and nibbles its coat to clean it (E). Barbary sheep may also take dust baths, lying down and using the tips of their horns to toss sand over their bodies.

descended from more than one wild species. Though it cannot be proven with certainty, the wild goat is the most likely ancestor of the domestic goat, found in rural areas throughout the world. Archaeological evidence suggests that domestic goats were in existence in the Middle East over 9000 years ago—predating the appearance of the first domestic cattle.

Relentless consumers

Domestic goats graze on grass and browse on leaves, twigs and buds, but they will eat almost anything if their usual food is scarce. When gathered in large numbers, goats devour all the plants in the vicinity. Unlike other ungulates, domestic goats and sheep nibble plants right down to the ground,

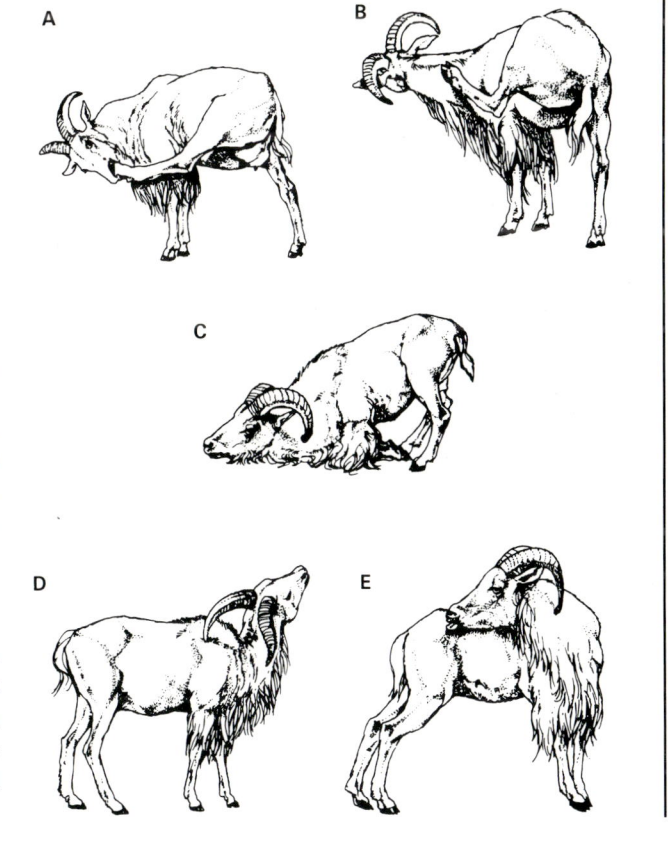

559

THE GOAT ANTELOPES — CAPRINS

ABOVE **The mouflon is on the verge of extinction in its original homelands, the Mediterranean islands of Cyprus, Corsica and Sardinia. The reddish-brown coat of the mouflon has a warm, woolly underfur in winter, protecting the animal from the bitter cold of its mountainous habitats.**

sometimes eating the roots too. The vegetation may then be killed off by frost or drought or be trampled by migrating herds. The tragic consequences can be seen in the vast areas of India and Africa that have been laid waste.

Feral goats (domesticated animals that have returned to the wild) have become the dominant species in many places. As a result, the native flora has often been entirely wiped out or brought dangerously close to extinction. The local fauna suffers too. Many birds (for example, on Hawaii) and insect species disappear, unable to find bushes in which to shelter or plants to eat, and wild ungulates struggle to survive in the face of competition for food and living space from domestic goats.

The Barbary sheep

Despite its name, the Barbary sheep (or aoudad) looks more like a goat than a typical wild sheep, but scientific tests show that it is related to both groups. Males measure between 60 and 65 in. from head to tail; standing up to 3 ft. 6 in. tall at the withers, they are much larger than females, and at up to 320 lbs. they are generally twice their weight.

The Barbary sheep's range once covered a large area in and around the Sahara, but now it survives in the mountains of North Africa (such as the Atlas range) and in the higher parts of the desert. Both sexes have curved horns (much larger in males) which are triangular in cross-section. They have neither the beard of a goat nor the rump patches of a wild sheep. Their coats are tawny-red in color and males boast a magnificent pale-colored mane of long, soft hairs that falls from the base of their necks and shoulders almost to the ground. Whitish hair covers the inside of the ears, the chin, a narrow stripe along the underparts and the inside of the legs.

Barbary sheep are most active in the early morning and evening. They eat roots and tubers in summer, lichen and grass in winter, and can survive on very little water if they have to—an important adaptation for a desert animal. Like other desert creatures, the Barbary sheep visits wet areas whenever the opportunity arises. Like goats and unlike sheep, they are not gregarious, but are either solitary or live in small groups—except in the mating season, which lasts from October to November, when they live in large mixed groups.

Trials of strength

Male Barbary sheep fight one another for the privilege of mating and show their strength with head-butting battles. They either stand head-to-tail and strike out at one another's hindquarters with their horns, or they interlock horns and pull in a tug of war. After the rut, males tend to wander far away from the females. Between March and May, following a five- to six-month gestation period, females give birth to one or two lambs (occasionally triplets). The young are quick to develop, following their mothers across the harsh terrain shortly after birth.

Barbary sheep have been hunted to extinction over most of their Saharan homelands. They have, however, been introduced to South Africa, the USA and Mexico where they are thriving, generally at the expense of indigenous species. In the USA, for example, they are competing for grazing with the bighorn sheep, which are already endangered in some parts of their range.

Blue sheep

The blue sheep, often called by its Hindi name, the bharal, lives in the Himalayas, and ranges from Kashmir through Tibet to Mongolia and eastern China. The animal's lack of a beard and pungent body odor suggest that it is more closely related to sheep than to goats. Nevertheless, the animal's horn

RIGHT A small group of mouflon march in a line along a narrow mountain path. Wild sheep often walk in single file when traveling any distance and bunch together only when danger threatens. Their hooves are widely splayed to spread their weight so that they do not sink into the snow.

structure and the sturdy projections produced by its second and fifth toes indicate that it may, in fact, be more closely related to goats.

Male blue sheep stand up to 3 ft. tall at the withers and weigh up to 180 lbs. Their handsome bluish-brown coats contrast with the white hair of their undersides, the area around the anal region and the inner surface of the legs. A prominent black band crosses their flanks between the bluish upper parts and the white undersides. Unlike goats, they lack knee calluses, and the tail is naked underneath. Males have long, heavy horns (up to 30 in. long) which curve backward and outward; females are much smaller than males and have small, pointed horns. Blue sheep live on open, grassy mountain slopes and plateaus, at altitudes of 8200-18,000 ft., where they feed on grasses, herbs and lichens. If approached by a predator, and there is no cover in which to hide, they remain motionless; but when they realize they have been spotted, they make a run for it, climbing the steep, inaccessible cliff faces.

Violent battles

During the rut, which lasts from October to December, males engage in violent head-butting battles. Keeping close to one another, they rear up on their hind legs, then fall forward to land a crashing butt before their forelegs touch the ground again—so the full force of their weight is behind each blow. After a dozen exchanges, one of the rivals is forced to recognize the superior strength of the other and abandons the fight. The winner chases the loser away and may strike him again in the flank or hindquarters.

The structure of the herd changes with the seasons. Males leave the females after the rut to form single-sex herds that may be 30-40 strong (although 5 to 20 is more usual). In April or May, after a five-and-a-half-month gestation period, females give birth to a single offspring. Twin births are rare. Although the blue sheep is not endangered, man is its main enemy. Its traditional predator, the snow leopard, has been wiped out over much of its range in the Himalayas by poaching for its skin.

Wild sheep

There are six species of goat antelopes referred to as wild sheep: the mouflon, urial, and argalis are closely related Eurasian sheep, while the snow sheep, Dall's sheep and bighorn sheep are distributed over North America and northeast Siberia. Sheep are gregarious animals, generally living in single-sex groups, except during the mating season. Unlike goats, their horns are massive and blunt, and curl to the side and forward.

THE GOAT ANTELOPES—CAPRINS

The adaptable mouflon

The mouflon is the only wild sheep in Europe. It is also the smallest, standing about 30 in. tall at the withers and weighing up to 110 lbs. Its coat is a dark chestnut-brown. The mouflon originated in Cyprus, Sardinia and Corsica, but since the 19th century has been introduced to many places in central and southern Europe and Asia Minor. Today there are hardly any mouflon surviving on their original island homes. In some places, the mouflon has been crossed with domestic sheep to enable this warmth-loving Mediterranean animal to withstand the bitter cold winters further north.

Once found at any altitude on the more gentle, open slopes of Cyprus, Sardinia and Corsica, mouflon have adapted to a wide range of habitats, from coniferous forest to deciduous woodland and to altitudes of over 5000 ft.

Females and their young live in large flocks all year round. Males form separate, smaller flocks, usually consisting of animals of similar age; some, especially the older rams, prefer to live on their own.

Clashes of will

Conflicts between males during courtship and at other times are resolved by head-butting and shoulder-to-shoulder combat. In the head-butt, the animal takes a short run before leaping at its opponent, meeting head-on with a crash of horns that can be heard echoing through the mountains. Less dramatic is shoulder-to-shoulder combat, in which each animal pushes and shoves the other while their horns are interlocked. Wild sheep do not normally display any territorial behavior, in spite of possessing preorbital, foot and groin glands for scent-marking strategic objects. They do, however, toss the head sideways if they need to threaten another animal.

Mating takes place during October (though it may occur later in harsher environments) when single males join the female groups and compete with other males for the females' attention.

From the end of February until the end of April, pregnant females go off on their own to give birth to one or two young. The young can run around soon after birth, but they need to be suckled every 15 minutes or so. When summer comes, females and their offspring (and males up to two years of age) re-group to form herds of around 30-40 individuals.

TOP The urial, the wild sheep of Asia, lives in the mountains from Iran to the Himalayas, with a small population in Oman discovered as recently as 1968. The male is easily identified by its massive horns, long white throat ruff and cheek beards.

ABOVE The argali lives in remote areas throughout the high plateaus of central Asia. The animal's name comes from the Mongolian word *arghali*, meaning simply "ram." The horns of this magnificent sheep grow to an average length of 70 in.

THE GOAT ANTELOPES—CAPRINS

RIGHT **A female Dall's sheep (center), accompanied by two youngsters whose horns are still relatively undeveloped, disappears over the brow of a hill. Outside the mating season, which begins in October and lasts through the winter, these sheep form small groups consisting of males only, or females and young.**

PAGES 564-565 **Female bighorn with their young feed in small groups during the spring and summer. They will be joined by the males as the mating season approaches. Two of the females have each lost a horn, either by falling on the steep mountainside or during a clash with another animal. The horns will eventually grow back to full size.**

Although primarily grazers, feeding mainly on grasses, mouflon, like goats, eat almost anything. They will even eat the tough, leathery plants that most other ungulates refuse to eat, enabling them to live in some extremely arid habitats.

The urial

The urial is closely related to the mouflon, but is a larger animal, weighing up to 190 lbs. Most urials have light brown fur, the shade and pattern of their coats varying from area to area. They inhabit undulating hilly country and arid plateaus in Iran, southern Soviet Central Asia, Afghanistan, Pakistan, Kashmir, northwest India and Tibet. An isolated population also lives in Oman. Though the range is extensive, the population is now extremely fragmented, and in many areas the animals are threatened. The main causes of their decline have come through competition for food with introduced domestic sheep, and through hunting for their meat and horns.

The argali

The largest of all the wild sheep, the argali may weigh as much as 400 lbs. It has a light brown coat, white legs and a large white patch on its rump. Like the urial, it is similar in form to the mouflon—but its horns are greater in size.

Today, argali are restricted to central Asia, though they may have been more widely distributed in prehistoric times. They range from the Altai and other mountain chains in southern Siberia through Mongolia, and Tibet to the Himalayas, spreading through the Tien Shan mountains of the USSR and China, and the Pamir range west to the mountain ranges of the Kyzyl Kum Desert in Soviet Uzbekistan.

Argali prefer open plateaus or gently rolling upland terrain to the steep rocky mountain areas inhabited by many goat antelopes. In most parts of their range, this type of country exists at high elevations—argali frequently live above 10,000 ft. and some subspecies are found as high as 19,000 ft., often near the limit of vegetation. Though ibex and argali occur in the Himalayas at similar altitudes, they do not compete for food because the ibex keep to the steep cliffs above the gentler valley slopes where the argali graze. The argali's habitat is remote and bleak, offering little shelter from the elements, and the animals usually descend to lower ground during harsh winter weather. To escape from predators such as wolves, they do not seek refuge among steep cliffs and rocks—a tactic used by American sheep and, sometimes, by the urial. Instead, they escape from danger by running.

Heavy horns

As with sheep in general, there is much controversy over the classification of the argali. Zoologists disagree about the number of subspecies, some recognizing seven, 14 or even 16, while others consider the number to be only four. One of the key differences between the various regional populations is the size of the animals' horns. The smallest of the subspecies,

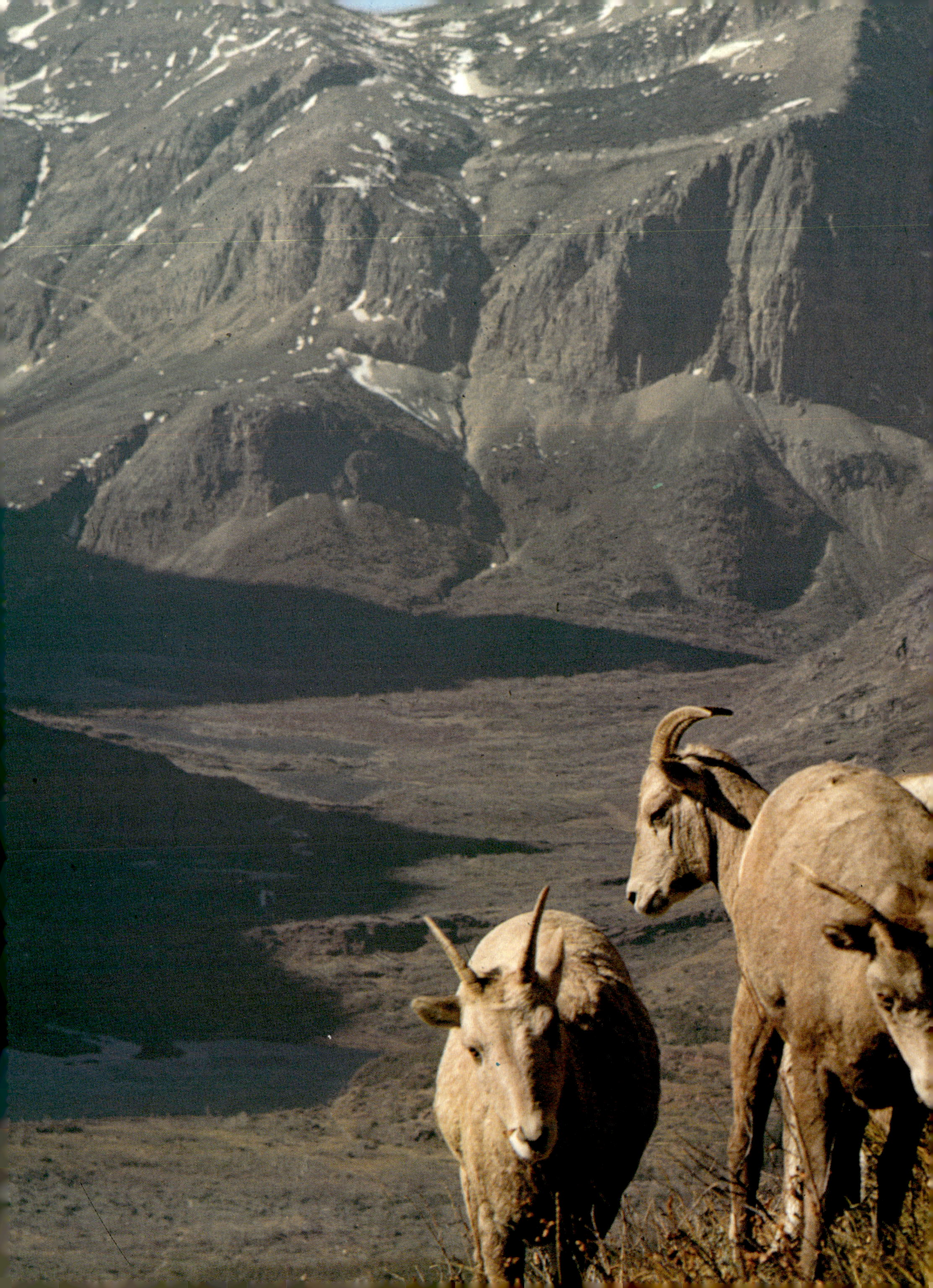

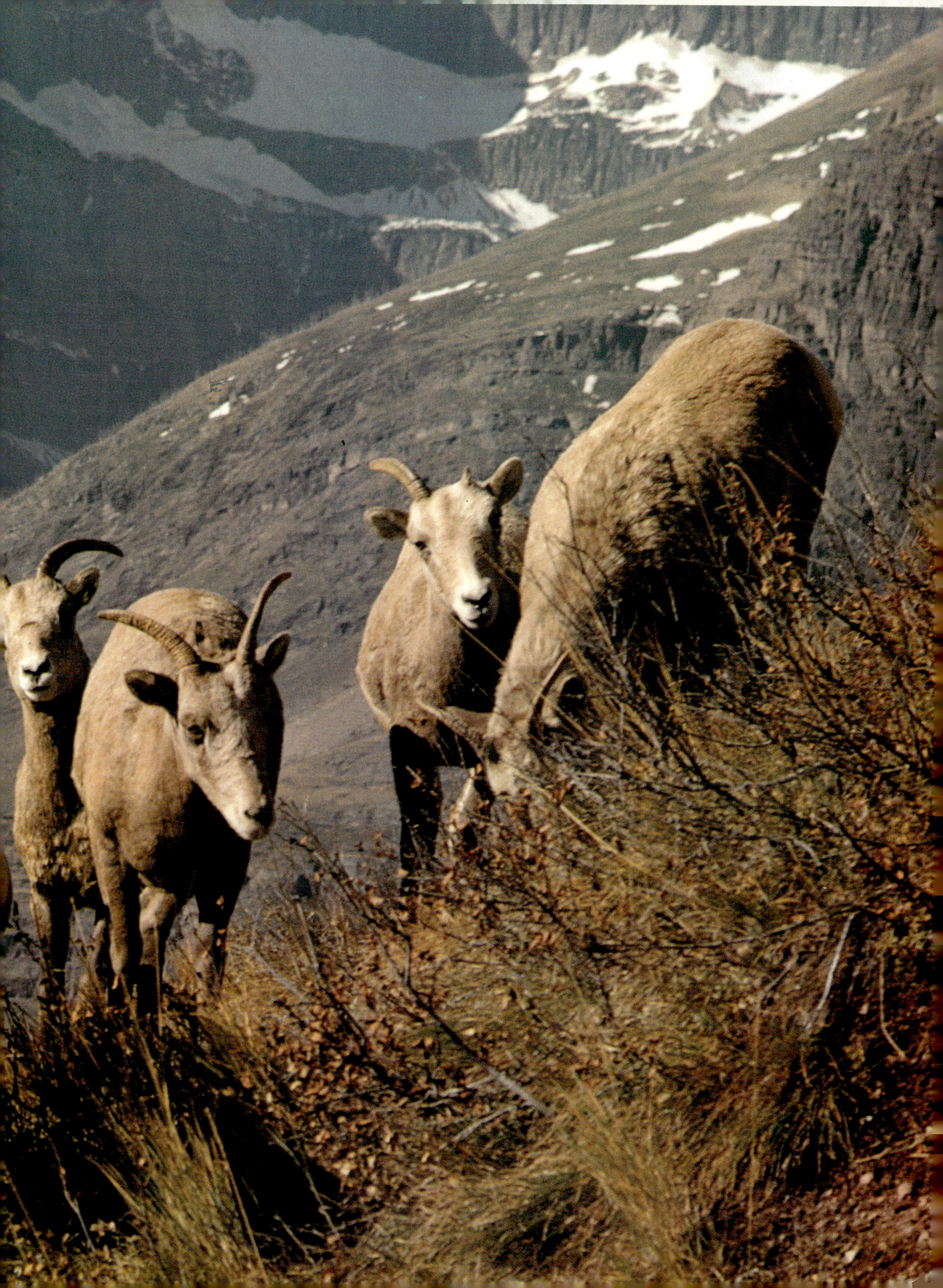

THE GOAT ANTELOPES—CAPRINS

Argali may not only suffer from lack of food during heavy winters, but they are also at risk from avalanches and wolf packs. However, without doubt their main enemy is now man. Their numbers have greatly declined in the 20th century because of hunting. Although they are legally protected in most areas, they are nevertheless extensively hunted. Consequently, argali are threatened over much of their range, and in some places they are thought to be in danger of extinction.

American sheep

Most zoologists regard the American sheep as three separate species—the snow sheep, Dall's sheep and the bighorn. The snow sheep in fact lives in the region of Siberia closest to North America, rather than on the North American continent itself. They are broadly similar in form to the Eurasian sheep, but show clear differences in behavior and in choice of habitat.

All three species of American sheep inhabit rugged as well as gentle mountain terrain. They may forage on high meadows, shrub-covered ravines or even cliff ledges. In the virtual absence of other goat antelopes (only the mountain goat shares the same highland regions of North America), they have partly taken over the terrain normally occupied by goats.

Snow sheep live in the coldest and highest regions of northeastern Siberia, especially in places where there are cliffs for shelter. The three subspecies of snow sheep (some zoologists believe there are four) are shorter and lighter than bighorn sheep, with slightly shorter horns. They have a coat of yellowish-gray hair, paler on the belly, with almost straw-colored hair on the neck and head, and some have a dark band across the muzzle.

Dall's sheep

Dall's sheep (sometimes called the thinhorn) lives in the cold, mountainous regions of the Northwest Territories, northern British Columbia, Yukon Territory and Alaska. It is slightly smaller and more lightly built than the bighorn (the male weighs 200-265 lbs.), with slimmer, more gracefully coiled horns. The color of Dall's sheep varies from light grayish-brown to dark brown, with a white tip to the muzzle and a large, white rump patch. Two of the three generally recognized subspecies, the common Dall's sheep and the Kenai Peninsula Dall's sheep, are pure white in color, while the black thinhorn sheep, also

TOP **A male bighorn ram curls back its upper lip after sniffing the urine of a female in heat—a facial expression known as the "flehman" response, displayed by many wild rams. To judge by the well-developed horns with their many rings, this is an old animal.**
ABOVE **Male bighorn rams spend much time fighting during the rut. Facing each other on their hind legs (broken-line drawings), they take a few steps forward before crashing their horns against one another, the full force of their weight behind the collision. The rounded horns do not cause injury, but when the animal is exhausted he will retire from battle, his place then being taken by another.**

Severtzov's argali (from the Kyzyl Kum Desert), has horns similar in size to those of the urial. At the other extreme, the Marco Polo sheep (an argali living mainly in the Pamir Mountains) has huge horns that form a long, open spiral, the tips twisting up, out and down again. Measured around the curves, the horns of one specimen reached 75 in. in length. Although the horns of the Marco Polo sheep are the longest, they are remarkably light in weight. Those of other argali are thicker at the base and much heavier. A ridge along the horns probably reduces the chance of them breaking during fights. In some subspecies the horns alone may weigh more than the animal's skeleton.

THE GOAT ANTELOPES—CAPRINS

known as Stone's sheep, varies from a silvery-gray to a glossy black, with white markings. An estimated 65,000-90,000 Dall's sheep survive today, the largest numbers in Alaska, where some populations appear to have recovered from earlier declines.

The bighorn

The range of the bighorn extends from the high mountains of northern Alaska, where temperatures plummet to below −58°F, to the arid lands of the southwestern USA and Mexico, including Death Valley, California, where the temperature soars to more than 122°F.

Bighorns are large sheep, almost as big as the argali, weighing up to 300 lbs. They are more compact and robust than Eurasian wild sheep, their skulls are stronger and shorter, and the bone that supports the horns is larger. Both sexes have tightly curled horns; those of the ram are far larger, growing up to 3 ft. 6 in. or more long and 18 in. or more in circumference.

The horns of a bighorn develop slowly and reach full size only when the animal is at least six-and-a-half years old (a mouflon's horns, by comparison, are almost fully grown when the animal is about four years old). The powerful horns and strongly armored skulls of mature bighorn rams are potent weapons when two rivals clash head-on in a fight to establish dominance.

In a similar way to the rutting behavior of red deer, when rival males gauge each other's relative strengths by roaring competitions, male bighorns are able to assess one another's strength through horn size. If a male is forced into an encounter with a group of unknown animals, he will attempt to pick an opponent with smaller horns. In this way, conflicts with much weaker or stronger animals are avoided. Males with the biggest horns mate most frequently, and these rams will inevitably be the oldest.

Human enemies

Being so strong and quick to move across the rocky mountain slopes, bighorns have few natural enemies. Pumas, wolves, coyotes and eagles occasionally kill a very young, sick or weak old animal, but it was the arrival of the first white settlers to western North America that brought disaster to this tough, wild sheep. Hunting, together with the introduction of domestic sheep, which competed successfully for

TOP Although horn growth is slow in bighorns, it is the main factor in deciding rank. Generally, bighorn males in a group follow the ram with the largest horns. The large, white rump patch is characteristic of bighorns; a subordinate animal turns the patch to face a dominant animal to indicate submission.
ABOVE Soay sheep have lived on St. Kilda, the westernmost islands of the Outer Hebrides, for a thousand years, and display much of the behavior of wild sheep.

567

HOOFED MAMMALS

There are over 200 species of hoofed mammals or ungulates. The diagram shows the links between the major groups, as they appear on pages 331-570 of the encyclopedia.

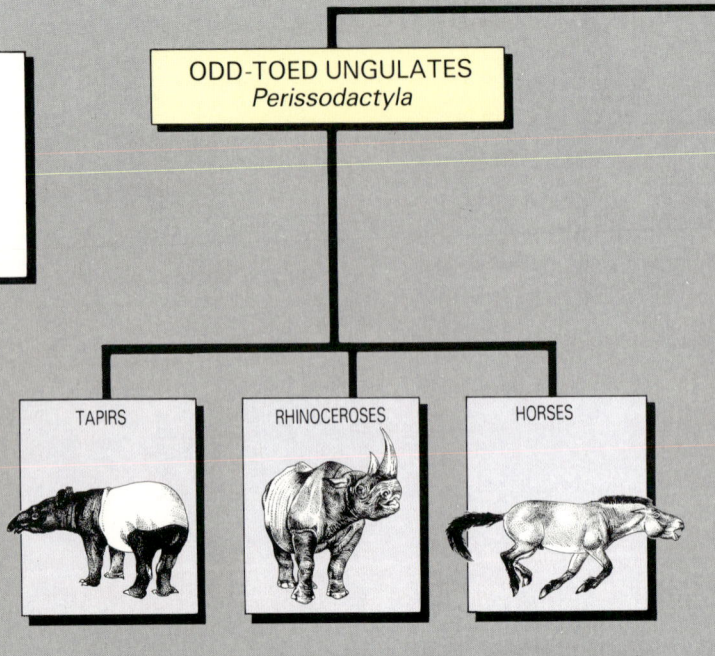

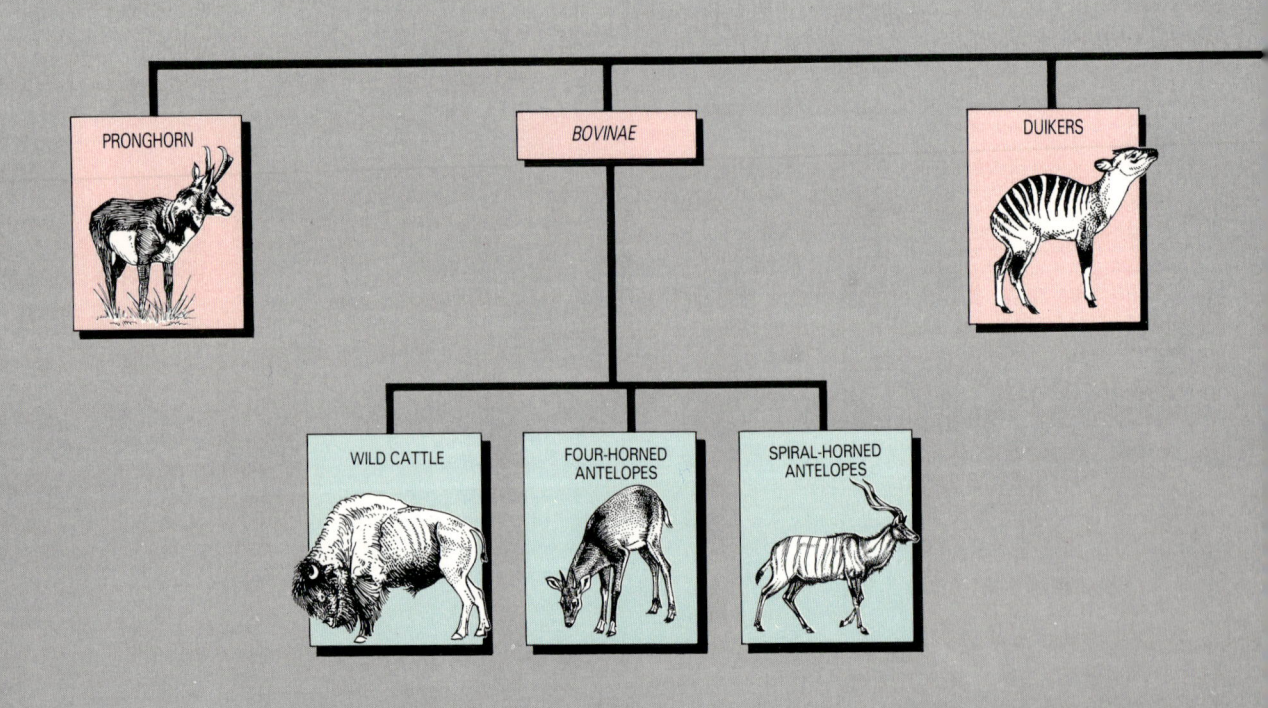

ODD-TOED UNGULATES
Perissodactyla

- TAPIRS
- RHINOCEROSES
- HORSES

- PRONGHORN
- *BOVINAE*
  - WILD CATTLE
  - FOUR-HORNED ANTELOPES
  - SPIRAL-HORNED ANTELOPES
- DUIKERS

THE UNGULATES

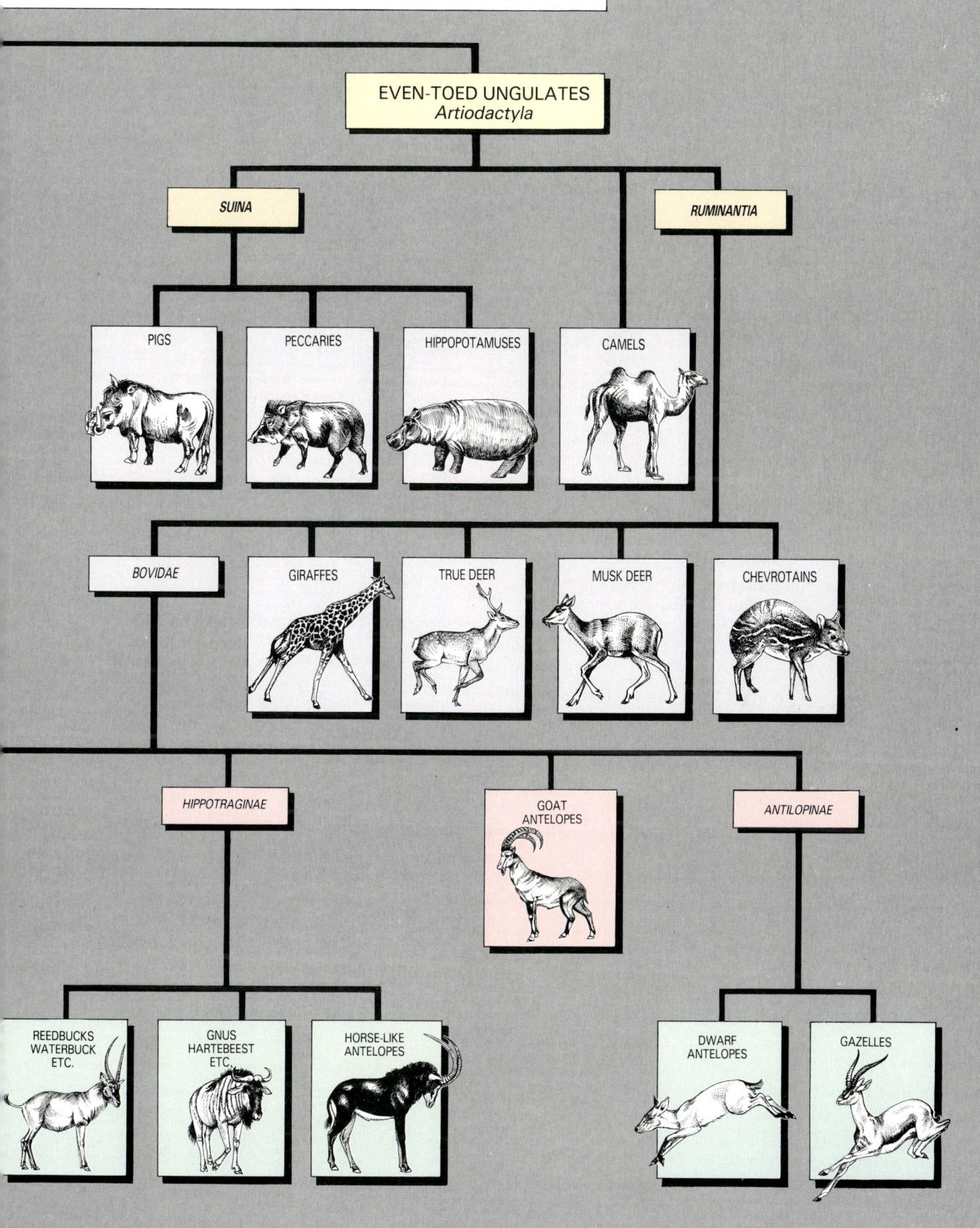

- **EVEN-TOED UNGULATES** *Artiodactyla*
  - *SUINA*
    - PIGS
    - PECCARIES
    - HIPPOPOTAMUSES
  - *RUMINANTIA*
    - CAMELS
    - BOVIDAE
    - GIRAFFES
    - TRUE DEER
    - MUSK DEER
    - CHEVROTAINS

- *HIPPOTRAGINAE*
  - REEDBUCKS WATERBUCK ETC.
  - GNUS HARTEBEEST ETC.
  - HORSE-LIKE ANTELOPES

- GOAT ANTELOPES

- *ANTILOPINAE*
  - DWARF ANTELOPES
  - GAZELLES

THE GOAT ANTELOPES—CAPRINS

GOAT ANTELOPES CLASSIFICATION: 3

Caprini

The tribe Caprini is the most advanced of the goat antelope tribes. The 17 species (grouped into five genera) include the animals most commonly referred to as goats, along with those known as wild sheep. Males are usually larger in size than the females and there is often a considerable difference in appearance between the two sexes. They range over Eurasia, North America and North Africa.

The genus *Hemitragus*, the tahrs, inhabit steep, craggy terrain in both arid and cold climates. There are three species: the Himalayan tahr, *H. jemlahicus*, of the Himalayan range; the Nilgiri tahr, *H. hylocrius*, of southern India; and the Arabian tahr, *H. jayakari*, which survives only in Oman in the Arabian Peninsula. Despite their names, both the Barbary sheep or aoudad, *Ammotragus lervia*, and the blue sheep or bharal, *Pseudois nayaur*, are closer in form to wild goats than to wild sheep. Barbary sheep live in the arid mountains of North Africa, while blue sheep are found in the Himalayas, Tibet and eastern China.

Six species belong to the genus *Capra*. All are confined to mountain regions, except the wild goat or bezoar, *C. aegagrus*, which inhabits a range of rocky, dry environments from the eastern Mediterranean through to India. The domestic goat, found worldwide, is most probably a descendant of this animal. The ibex, *C. ibex*, ranges over the mountains of central Europe, the Middle East and central Asia, while the Spanish goat or Spanish ibex, *C. pyrenaica*, is confined to the Iberian Peninsula. The markhor, *C. falconeri*, of central Asia inhabits mountain slopes at lower altitudes than the ibex. Two species occur in the Caucasus mountains between the Black Sea and the Caspian Sea in the USSR—the East Caucasian tur, *C. cylindricornis*, and the West Caucasian tur, *C. caucasia*.

The classification of the genus *Ovis*, the wild sheep, is a subject of much debate among zoologists. Some consider there to be only two species in the genus, others believe there are a total of six species. Three Eurasian sheep are very closely related: the mouflon, *O. musimon*, which lives in a range of habitats in central and southern Europe, Turkey and Iran; the urial, *O. orientalis*, of central Asia; and the argali, *O. ammon*, which ranges from central Asia to Mongolia. The three remaining species are also very closely related: the snow sheep or Siberian bighorn, *O. nivicola*, of northeastern Siberia; Dall's sheep or thinhorn, *O. dalli*, of Alaska and western Canada; and the bighorn or mountain sheep, *O. canadensis*, which ranges from Alaska to northern Mexico.

grazing and spread disease, drastically reduced the numbers of bighorn sheep. From a million or more in the mid-19th century, they now number only 35,000-40,000. Some populations, however, are thriving in protected areas such as the Yellowstone and Glacier National Parks.

Domestic sheep

The domestic sheep is the most common and most widespread of all ungulates. It is not certain which type of wild sheep gave rise to this hardy, domestic creature, but they are thought to have been first domesticated in the eastern Mediterranean region as much as 11,000 years ago, probably from the mouflon or one of its ancestors. Sheep give milk, wool and meat and need little attention compared to cattle; they also survive in areas where there is precious little to eat. There may be between 800-900 million sheep throughout the world. In Australia, sheep farming is big business, and the animals manage to live in semi-arid regions baked by the fierce sun.

Through selective breeding, humans have produced over 800 breeds of domestic sheep. Each breed has its own special qualities. Karakul sheep, for example, produce some of the finest wool (the well-known astrakhan fur is the fleece of their newborn lambs). Merino sheep, originally reared in Spain and bred for their wool, are now found in many countries the world over, and have served as important foundation stock for other breeds. Sheep sometimes return to the wild, but they do so much less frequently than domestic goats.

WHALES AND DOLPHINS—CETACEANS

TITANS OF THE DEEP

Ranging over the oceans of the world, the great whales and dolphins—some of the former weighing up to 170 tons—include remarkable acrobats and skilled communicators

WHALES AND DOLPHINS—CETACEANS

The whales and dolphins are mammals perfectly adapted to an aquatic life. They range in size from the porpoises, some of which are shorter than human adults, to the great whales, giants larger even than the prehistoric dinosaurs. Sadly, many species are now scarce, due to relentless hunting for their meat and oil.

Whales and dolphins are so completely adapted to water that they cannot survive on dry land. Whereas walruses and seals regularly haul themselves onto rocks and beaches to rest and breed, whales and dolphins invariably die if they are stranded out of water. They are unable to breathe because their own body weight crushes their lungs when it is no longer supported by surrounding water. Though they are generally hairless, and though some species possess shark-like fins, they are in every sense mammals: they breathe using lungs, they are warm-blooded and they suckle their offspring on breast milk.

BELOW Helpless sperm whales stranded on a beach make a tragic sight. The reasons for such mass strandings are unclear, but it has been suggested that the whales' sensitive navigation systems may be confused by sandbanks and even by magnetic fields.

PAGE 571 Easily recognized by its striking markings, a killer whale leaps clear of the water. The killer whale is the largest of the dolphins—some individuals grow to about the length of a bus—yet it is still dwarfed by the great baleen whales.

BELOW The traditional migratory routes of the blue whale (continuous lines) and the fin whale (broken lines) around the northern polar region.

BOTTOM The migratory routes and the main winter breeding grounds of the humpback whale in the oceans of the Southern Hemisphere.

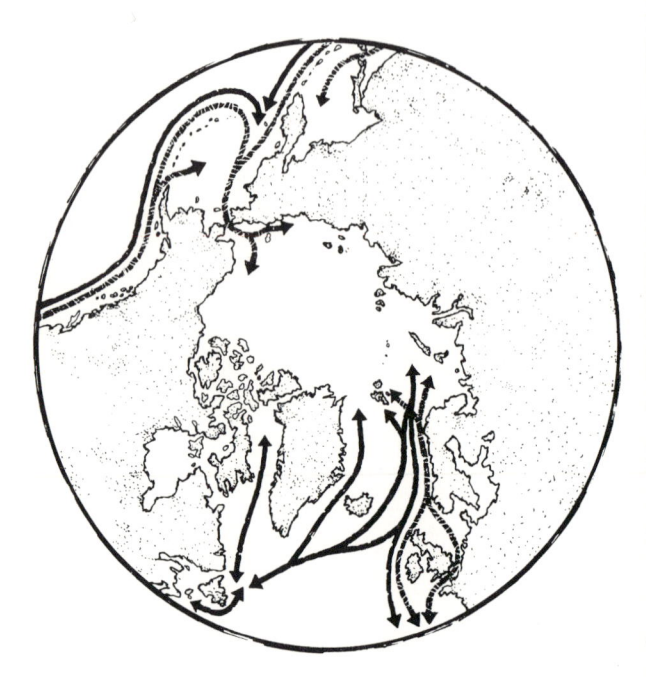

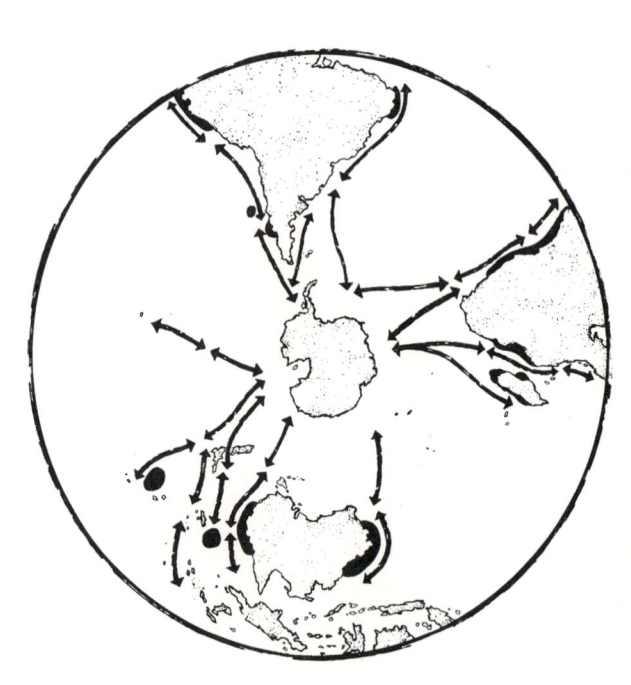

WHALES AND DOLPHINS—CETACEANS

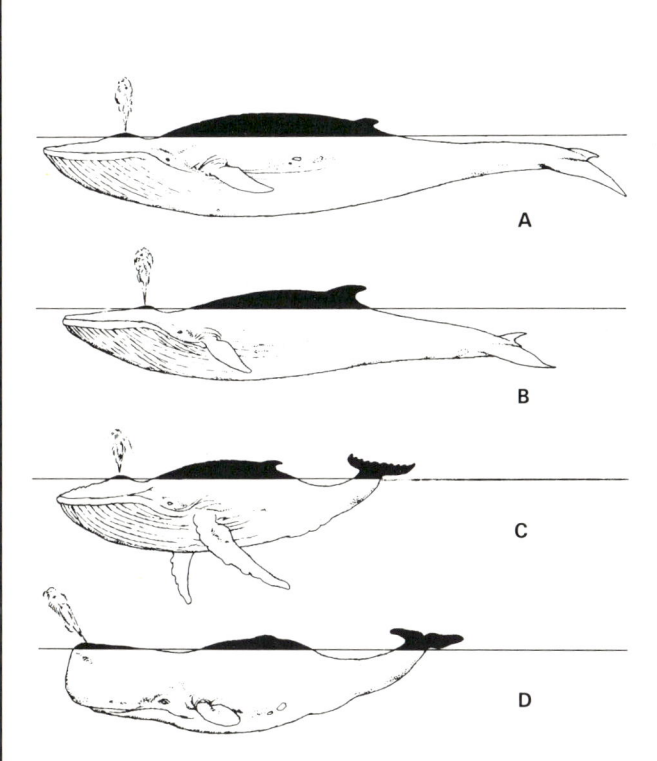

Evolution of the whales

In the early 1980s, fossilized remains of the most ancient whales yet known were discovered in Pakistan. The fossils were some 50 million years old, and the animals they came from still bore traces of a terrestrial ancestry. Their elongated snouts, for example, had nostrils at the front end, like land mammals, rather than on top, as in modern whales. The skulls of these early whales resembled those of certain animals within the prehistoric order known as the Condylartha, suggesting that the whales share a distant link with the primitive ungulates and the hoofed mammals.

The early whales may have been amphibious, spending most of their time in the water, but coming out on land to breed. Between 38 and 25 million years ago, most of the special features of truly aquatic, modern whales evolved. The changes included a streamlining of the body into a torpedo shape for fast, efficient swimming and the loss of external hind limbs (modern whales still have the bony remnants of the pelvic girdle within their bodies).

Though the rear limbs have disappeared, the front limbs of whales and dolphins have remained, albeit in greatly modified form. The arm and forearm bones are reduced in size while the toes have become enlarged, flattened "fingers" that support horizontal

ABOVE The tails of many whales break the surface when they dive. Like all whales and dolphins, but unlike fish, this killer whale has horizontal tail fins, or flukes, that move up and down to propel the animal through the water. The tails of fish are vertical and are waved from side to side.
ABOVE LEFT Each species of whale sends out a distinctive plume of spray, or "spout," from its blowhole when it reaches the surface. The pattern made by the whale's back, dorsal fin and tail is also a clue to identification. The drawings show a blue whale (A); a fin whale (B); a humpback whale (C); and a sperm whale (D). Note the unusual angle of the spout produced by the sperm whale.

paddle-like flippers. The flippers in whales and dolphins do not provide power; they are used for stability and for controlling direction through the water. Propulsion comes from the tail; unlike the flippers this is a boneless fin, or fluke. It is flattened horizontally, and powered by two large groups of muscles, one above and one below the base of the tail, that move it up and down.

The whale's skin is particularly well adapted to life in the water. It is smooth and hairless, except for a few places around the head. Hair would create turbulence in water, impeding the whale's movement through the sea. Layers of fat beneath the skin, called blubber,

573

WHALES AND DOLPHINS—CETACEANS

insulate the animal so that it can survive in the coldest oceans. In the bowhead whale of the cold Arctic region the blubber may be up to 10 in. thick.

Like all mammals, whales and dolphins have lungs. So, unlike fish that use their gills to extract oxygen from the water, they have to come to the surface to breathe. A whale's lungs are small in relation to its large size, and when the animal surfaces, it forces the stale air out from its lungs very quickly, rapidly refilling them with fresh, oxygen-rich air. In one breath, whales can renew almost 90 percent of their lung contents, compared to a figure of 10-15 percent for man.

The spout

When they return to the surface, whales and dolphins can empty their lungs completely in just one or two breaths through their blowhole (or blowholes). The sudden and tremendously powerful exhalation of stale air stored up during the animal's long dive is known as the "blow" or "spout," and its plume of spray can be seen for miles. Recent research indicates that the spout may consist of secretions of oil droplets from the glands lining the windpipe.

Each species of whale has a characteristic spout. Whales can, therefore, be sighted and hunted without one species being mistaken for another. The right whale, for example, has a double spout, the rorqual a single one, while the sperm whale's spout shoots out at a 45 degree angle. Whalers of the past had a skilled lookout, trained to identify the whales from a distance, and he enjoyed a rank almost equal to that of the helmsman or harpoonist.

The human predator

Whales have a long life span, a long gestation period and a low birthrate. In addition, they have few natural predators (mainly killer whales), which means that their slow population growth rate is not naturally disturbed. However, man, by hunting whales on such an extensive scale, has made it impossible for them to increase fast enough to make up their losses. Today, even if a total ban on whaling were imposed and fully observed by all countries, it would take a very long time for some species to recover to healthy levels. Indeed, some may never recover.

Whaling is an ancient activity. Whale bones have been found in the refuse heaps of Alaskan Eskimos and other native Arctic peoples from at least 6000

TOP **The killer whale has large, powerful teeth for grasping big fish, squid and other marine animals. Most whale species are "toothed whales," including the dolphins, the porpoises, the white whale, the sperm whales and the beaked whales.**

ABOVE **The baleen (whalebone) plates from the mouth of a blue whale. The rorquals, the gray whale, and the right whales all filter their food—plankton and other small marine animals—through the fine filaments that make up the baleen.**

574

WHALES AND DOLPHINS—CETACEANS

years ago. The French and Spanish Basques, the first major European whalers, started hunting right whales in the Bay of Biscay before the 12th century. The right whale was literally the "right" whale to catch: it was comparatively slow and continued to float after it had been killed, unlike other species which sank.

"Thar she blows!"

From the early 19th century, the sperm whale became the prime object of whalers. The sperm whale oil, when burned, gave better illumination than the oil of other species, while the waxy substance (spermaceti) found in its head made the best candles. By 1846—the peak year for whaling—some 729 American whalers were scouring the vast Pacific for sperm whales. At the same time, they took to massacring southern right whales and humpback whales whenever the opportunity arose. "Thar she blows" became the famous cry of the lookout when he spotted the whale's give-away spout.

In the 1860s, the invention of harpoon guns with explosive heads, by the Norwegian captain Svend Foyn, launched the age of truly merciless killing. The explosive harpoon now allowed hunters to attack those whales which had been too fast to hunt. Steam ships were used in the chase, and factory ships stayed

TOP Several recently killed sperm whales lie ready for processing near the stern of a factory ship. Today petroleum and electricity have long since replaced whale oil as a fuel for lighting, but Japan and Indonesia still hunt sperm whales for their meat.

ABOVE A leaping humpback whale displays its extremely long, pale flippers which alone may be 16 feet in length. It is also identifiable by its slender head with its flattened top, bearing numerous fleshy knobs or tubercles.

WHALES AND DOLPHINS—CETACEANS

ABOVE The head of a southern right whale bears large callosities (hardened patches of skin) that become infested with colonies of parasites, including barnacles, lice and parasitic worms. Right whales always have a prominent callosity, known as a "bonnet," on the top of their snout. Gray whales and humpbacks have the same kind of infestation, but the callosities are spread over their bodies, rather than clustered on their heads.

out at sea for long periods processing the dead giants. The whales were slaughtered in vast numbers. In December 1946, the International Whaling Commission was set up to control where and in what number the whales were being slaughtered. But despite this, whales are still being killed and some populations are at dangerously low levels.

Now another problem faces the whales—pollution of the world's oceans by oil, pesticides, waste products from industry and radioactive waste. Among the most polluted areas in which whales and dolphins are likely to be seriously affected are the Baltic Sea, the southern North Sea, the Bay of Fundy (around Nova Scotia) and the Gulf of Mexico.

Toothless monsters

Some of the baleen whales are the largest animals ever to have lived, far bulkier than even the biggest dinosaurs. The blue whale (one of the rorquals) is largest of them all; it grows to 100 ft. long and may weigh the equivalent of 25 African bull elephants (four times heavier than the largest known dinosaur).

Only in water is it possible to grow to such an extreme size. On land, bones could not support such bulk—they would simply break—but the bones of a blue whale serve only to anchor the muscles. The water gives the body the support it needs. Being so large has its advantages: because the surface area through which heat is lost is relatively small in relation to the overall volume of the animal, it is able to maintain a steady body temperature. Consequently, baleen whales are able to tolerate both the freezing cold of polar seas and the warmer waters of the tropics.

Mighty feeders

The heads of baleen whales are extremely large, sometimes making up over a third of their body size. Instead of teeth, these whales have long, slender

WHALES AND DOLPHINS—CETACEANS

The distribution of the right whales is split into two bands—one in the northern oceans, one in the southern.

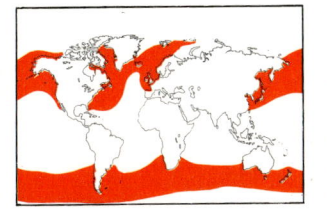

RIGHT When seen from above, the massive blue whale has a streamlined shape that enables it to swim fast, sustaining speeds of over 20 mi. per hour for up to 10 minutes or more. Its speed, and the fact that its body sank as soon as it was killed, saved the blue whale from being overhunted until the mid-19th century. But the arrival of modern whaling methods (such as the use of powerful harpoons and factory ships) rapidly brought the largest animal in the world to the brink of extinction.

baleen plates hanging from the upper jaw. The baleen is commonly known as "whalebone"—but is not actually bone at all. It is a substance similar to human fingernails, called keratin. The plates bear filaments on the inner side of the mouth forming a sort of fringe or filter. These collect vast amounts of plankton (tiny organisms) suspended in the water. All baleen whales eat by a process of filter-feeding, but the methods they use differ slightly in the various species.

To feed, the whale holds open its huge mouth, scooping in tons of seawater containing plankton (also krill and fish, depending on the whale species). As the whale raises its tongue and lower jaw to close its mouth, the water is forced out through the baleen plates, leaving the plankton caught in the baleen's bristly fringe. The whale then uses its tongue to remove the plankton before swallowing it. The success of this way of feeding depends on there being enormous amounts of plankton and krill in the baleen whales' rich polar feeding grounds.

Taking a deep breath

The bowhead whale (also called the Greenland right whale) lacks a dorsal fin, and is recognized by its enormous head (about 40 percent of its body length), greatly arched lower jaw and white chin on a black body. Bowheads weigh up to 80 tons and normally measure 50-60 ft. from head to tail. They are slow swimmers but can remain underwater without taking a breath for up to 45 minutes. Because of their thick blubber and exceptionally long baleen (10 feet or more), bowheads were considered by whalers to be one of the most valuable whales to catch. As a result, the number of bowheads plummeted from a mid-19th century population of 20,000-30,000, to the estimated 2000-3000 that survive today, making the bowhead one of the most endangered of all whales. Strict protection of the species is essential.

BALEEN WHALES CLASSIFICATION

Whales and dolphins form the order Cetacea. The order is divided into two groups—the members of one group have teeth, while the others possess baleen plates. The toothed whales make up the suborder Odontoceti, while the baleen whales form the suborder Mysticeti.

There are ten species of baleen whales in three families. The rorquals, of the family Balaenopteridae, include the humpback whale, *Megaptera novaeangliae*, the blue whale, *Balaenoptera musculus*, the fin whale or common rorqual, *B. physalus*, the sei whale, *B. borealis*, and the minke whale, *B. acutorostrata*, all of which range from polar to tropical seas. Another rorqual, Bryde's whale, *B. edeni*, occurs in both tropical and subtropical seas.

The family Balaenidae, the right whales, includes the pygmy right whale, *Caperea marginata*, which lives in the southern oceans; the bowhead or Greenland right whale, *Balaena mysticetus*, of the northern polar region; and the right whale, *Balaena glacialis*, which occurs mainly in temperate waters in both hemispheres. (Many zoologists consider the right whale of the Southern Hemisphere to be a separate species, *B. australis*.) Only one species belongs to the family Eschrichtidae—the gray whale or devilfish, *Eschrichtius robustus*, which lives mainly in coastal waters in the northern Pacific Ocean.

577

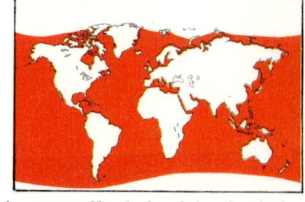

WHALES AND DOLPHINS—CETACEANS

The rorquals occur in coastal shallows and in deep seas throughout the world.

The right whale is sometimes called the black right whale because its body is mainly black, although it may have white patches on its belly. Its head and jaw have large skin callosities (thickened patches of skin), nearly always infested with parasites. Right whales normally measure around 50 ft. long and weigh about 55 tons, while large females may weigh up to 100 tons. Right whales live in the temperate waters of both hemispheres, in small groups containing up to a dozen whales. There are an estimated 4000 right whales surviving in the oceans—some 500-1000 in northern waters and about 3000 in southern waters.

Coastal giants

Gray whales prefer coastal waters and are often seen within about a mile of the shore. They measure 35-50 ft. long. The female weighs 33-38 tons and is twice as heavy as the male. Instead of a dorsal fin, the

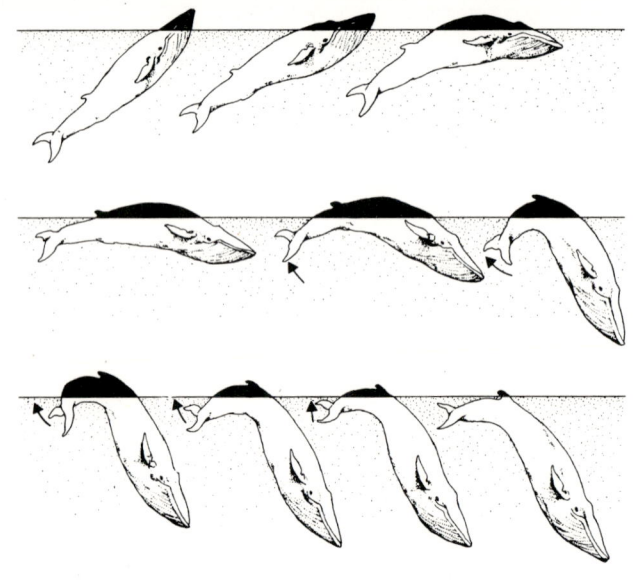

TOP A bowhead or Greenland right whale comes up for air. Its head comprises two fifths of its total body length, and it has a huge mouth. Moreover, it has very thick blubber and the longest baleen plates of any whale —features that have made the species all too attractive to hunters.
ABOVE As a rorqual comes up to breathe, the whole length of its back breaks the surface, and the tail arches to take the animal underwater again.

THE GREAT WHALES

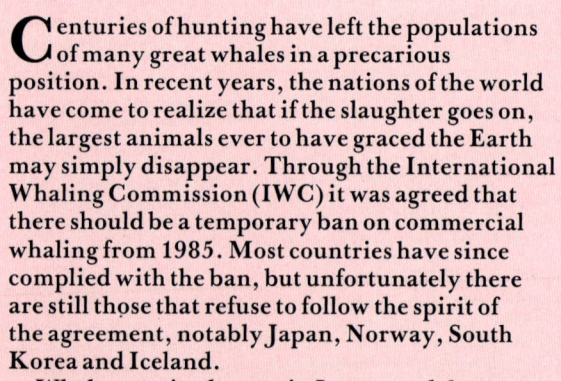

Centuries of hunting have left the populations of many great whales in a precarious position. In recent years, the nations of the world have come to realize that if the slaughter goes on, the largest animals ever to have graced the Earth may simply disappear. Through the International Whaling Commission (IWC) it was agreed that there should be a temporary ban on commercial whaling from 1985. Most countries have since complied with the ban, but unfortunately there are still those that refuse to follow the spirit of the agreement, notably Japan, Norway, South Korea and Iceland.

Whale meat is a luxury in Japan, and that country continues to provide the greatest demand for whale products. As long as the demand exists, hunters will always try to find a way to meet it, and the IWC has no effective power to stop them. By the end of 1987 at least 11,000 whales had been killed since the start of the ban. Moreover, the whalers have been quick to exploit a loophole in the international agreement, which permits whaling on the grounds of scientific research. In 1986 Icelandic hunters took 120 whales from the sea, claiming that the killings were carried out in order to assess population levels. Like before, much of the whale meat was then exported to Japan.

WHALES AND DOLPHINS—CETACEANS

whales possess a low "corrugated" ridge along the last section of their back. As their name suggests, gray whales are grayish in color and are covered with elaborate patterns formed by barnacles which live on their skin. Gray whales are seabed feeders, sweeping through the sediment for crustaceans, worms and mollusks, which they filter with their short baleen plates.

Until the early 1700s, there were three separate populations of gray whale: the North Atlantic, the Californian and the Korean. Today, with the North Atlantic species extinct through over-hunting, only the last two remain. They are especially famous for their great migrations from the summer feeding areas in the sub-Arctic to the winter calving areas in the subtropics.

Marathon migrators

The two stocks have different migration routes. The Korean gray whales (which may be close to extinction) travel from the Sea of Okhotsk, bordering eastern Siberia, down to the coast of South Korea. During the summer months, the Californian animals take full advantage of the rich feeding grounds off the coast of Alaska in the northern Bering Sea and the Chukchi Sea that separate Alaska from Siberia. When they migrate south, they keep close to the North American coast, swimming through the Gulf of Alaska, along the coast of British Columbia, Oregon and California, to the winter breeding grounds off the coast of Baja California, Mexico.

Not all whales migrate the full distance, but it is nonetheless an incredible journey of up to 6400 miles (one way). It is not clear how they navigate but, since they keep close to the coast at all times, it seems likely that they know land is on their left as they swim south and on their right as they travel north.

The rorquals

The rorqual family include the real giants of the whale family, such as the blue and fin whales. They all have dorsal fins, and furrows under the throat which allow it to expand when feeding. Many rorquals travel enormous distances on annual migrations from tropical waters to the polar regions. Their baleen plates collect vast amounts of krill and plankton, sometimes even shoaling fish such as herring.

The blue whale is the largest animal ever to have lived, with dimensions that stagger the imagination. It grows to between 75 and 90 feet in length and weighs about 165 tons. The blue whale's heart alone weighs 1000 lbs., its liver 2200 lbs. and its tongue 6600 lbs. (the weight of a full-grown hippopotamus). The skin is a mottled blue-gray and the undersides of the flippers are pale. It may live to 80 years old, and its range extends across all the oceans of the world from tropical to polar waters. Its subspecies, the 75-ton Pygmy blue whale, is silvery gray in color and confined to the Southern Hemisphere.

TOP **A gray whale at its winter breeding grounds off the coast of Baja California in Mexico. Reserves of fat built up during the summer are sufficient to sustain the whales in the winter.**

ABOVE **The minke whale is the smallest of the rorqual family and has a white band on its flippers. As populations of larger whales decrease, the minke whale has become the victim of hunters.**

WHALES AND DOLPHINS—CETACEANS

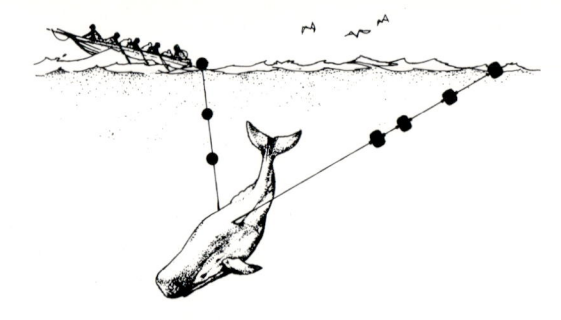

TOP A group of female sperm whales and their young. Females are not as well insulated against the cold as male sperm whales, and do not venture with them into polar waters.

ABOVE In the days before modern whaling, sperm whales were harpooned from rowboats. The whalers used lines fitted with floats of sealskin (left), and more buoyant floats of cork (right).

Before commercial whaling started there may have been as many as 230,000 blue whales throughout the oceans. By the time they received legal protection in 1965, probably as few as 7000 remained. Today, despite the protection, their numbers show no signs of recovery. Unfortunately, blue whales have a low rate of reproduction. Only one offspring is born at a time, after developing for some 11 months in the womb. The young are suckled for at least seven months and their mothers will not give birth again for another two years. If there is a plentiful supply of unpolluted food, a mother may produce 10 offspring in her lifetime.

Slender relative

The second-largest whale is the fin whale, reaching some 85 ft. in length. Its other name, common rorqual, is something of a misnomer, for its population has

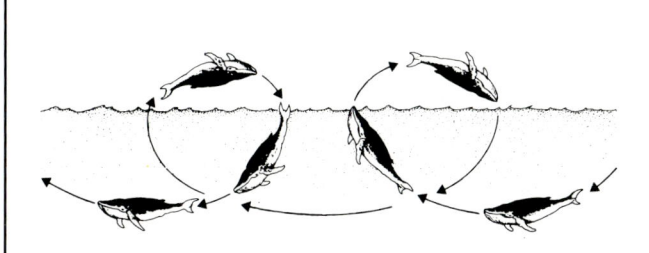

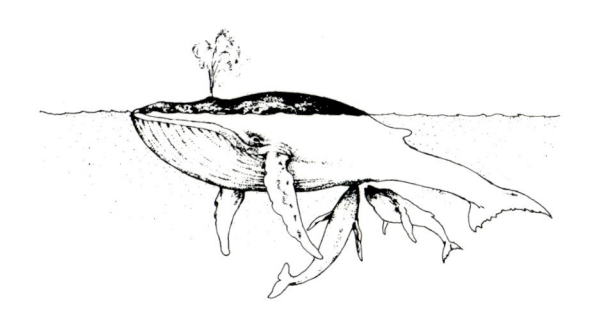

WHALES AND DOLPHINS—CETACEANS

TOP AND TOP RIGHT Despite their considerable bulk, humpback whales are well known for the spectacular leaps that often take them right out of the water. Known as "breaching," such acrobatic behavior may be both a form of communication with other whales, and a means of stunning and rounding up shoals of fish.
ABOVE Female humpbacks stay on the water's surface to suckle their young. They suckle for up to about 12 months before the offspring is weaned. At birth, their offspring are already over 13 ft. long.
RIGHT Growing to no more than 11 ft. in length, the pygmy sperm whale is tiny compared to its enormous relative, the sperm whale.

dropped from about 500,000 in the days before whaling, to a current estimate of 100,000. Though it is only a few yards shorter than the blue whale, it has a much lighter build, and rarely exceeds 75 tons. Its upperside is gray and it is whitish underneath, with white undersides to its flippers and tail. The fin whale is thought to be one of the fastest swimmers among the great whales, and its range extends from polar to tropical seas. More gregarious than most other rorquals, it may gather in groups of up to 100 in plankton-rich feeding grounds.

The minke whale is the smallest member of the rorqual family. It is around 32 ft. long and weighs approximately 10 tons. It lives in polar and tropical waters in both hemispheres, the northern populations having a distinctive white band on the upper surface of the flippers. Minke whales often occur in small groups and may gather in large herds in places where food is especially abundant. South of the Equator, the animals benefit from the greater supply of food created by the slaughter of larger rorquals. They presently number about 500,000, but are themselves now being hunted by whalers. They are also hunted by killer whales.

The humpback whale is a member of the rorqual family, though it belongs to a different genus from the other rorquals. Adult humpbacks are about 50 ft. long and weigh around 50 tons, the females growing slightly larger than the males. They almost always have

581

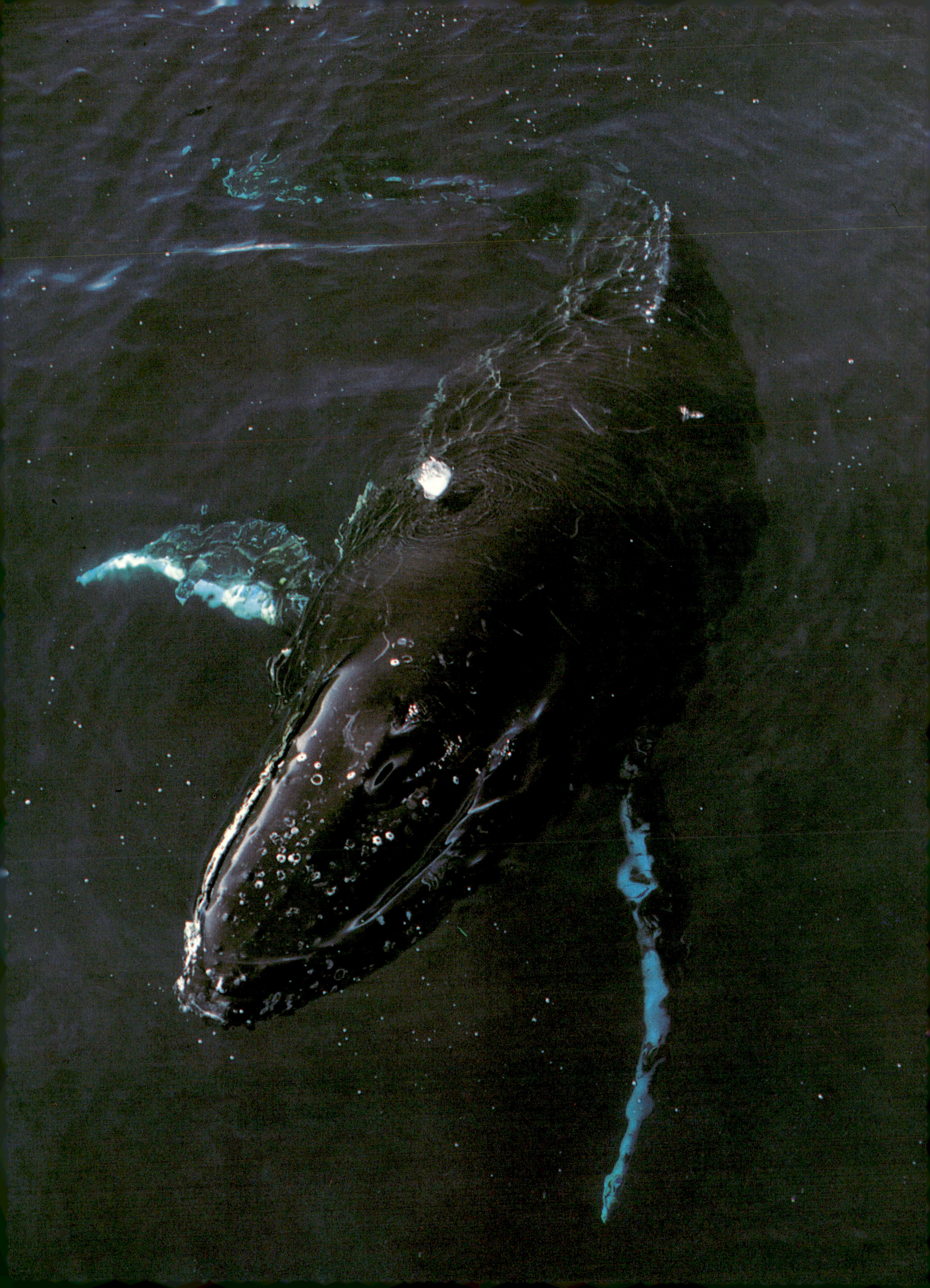

WHALES AND DOLPHINS—CETACEANS

LEFT **Humpback whales produce eerie underwater songs that can carry for great distances through the ocean. Their repertoire includes squeaks, sighs, bellows and haunting groans.** RIGHT **A surfacing gray whale exhibits the multitude of barnacles that encrust its body,** giving it a mottled appearance. **One species of barnacle lives only on gray whales. Along the coast of Siberia, gray whales have been observed taking showers in the cold, fresh water pouring from cliffs, possibly in order to drive off their numerous skin parasites.**

patches of barnacles on their skin, forming striking patterns. The best way to distinguish humpbacks from other rorquals is by their enormous flippers, which are almost one third of their total body length. Like other rorquals, humpback whales have a series of throat grooves that allow their throats to expand considerably when they feed. Humpbacks live in polar and tropical seas, both north and south of the Equator. Despite their size, they frequently perform acrobatic leaps at the surface, often clearing the water completely and returning to it with a tremendous splash.

The "singing" humpback

The humpback whale is famous for its "song." The strange series of calls can last from five to 35 minutes and consists of a wide variety of sounds, complex in structure, with many patterns recurring through the song. The sounds are so penetrating that they have been detected over 110 miles away. Solitary males in coastal breeding waters probably sing in order to advertise their presence to potential partners. Since each whale has a distinctive voice of its own, it is possible that females can identify the singer.

Although an individual's singing performance can be distinguished from that of other humpbacks, all the whales in the same breeding ground sing the same basic song. The construction of the song varies from season to season, certain phrases being dropped, others added, some merely altered slightly. The noise from ships and outboard motors is a serious disturbance to this form of undersea communication, and may obstruct the mating habits of the whales. If so, it could be an obstacle to the recovery of the species, following years of severe hunting pressure. Before whaling began there was probably a total of some 115,000 humpbacks in the world—some estimates put their present number as low as 4000-6000.

Broad tastes

Though they are all baleen whales, the rorquals have differing tastes in food. Some, like the blue whale, feed almost exclusively on shrimp-like organisms, while others rely heavily on fish. Fish swallowed by whales are usually shoaling species such as herring, mackerel and cod. They also feed on squid when they find them in large concentrations. To catch their food, rorquals may quarter an area back and forth, forcing their prey close to the surface, before engulfing them in their open mouths. Humpbacks may also herd fish by diving after them, and then lunging at the fleeing shoal. Following examinations of the stomach contents

WHALES AND DOLPHINS—CETACEANS

LEFT Like all river dolphins, the Ganges dolphin has very poor eyesight. In its muddy environment—the Ganges river in India—clear visibility is limited to no more than a few inches. The animal relies, instead, on echolocation to locate its prey and its surroundings.

The toothed whales

Most species within the whales and dolphins group possess teeth rather than plates of baleen. They include the sperm whale, beaked whales, river dolphins, white whales, porpoises and the dolphins.

The sperm whale is the giant among the toothed whales, quite different in shape from the baleen whales. Males grow up to 60 ft. long and weigh up to 80 tons. Its toothed lower jaw is short and narrow, and it has an enormous, square-shaped head constituting about one-third of its total body length. Its blowhole is located at the front of its head, and slightly to one side, producing a distinctive spout that is expelled at a 45 degree angle.

Underwater drama

Sperm whales mainly eat squid and octopus but will also prey on rays and dogfish, as well as smaller fish and crustaceans. They have a particular liking for giant squid, which can measure up to 50 ft. in length. These monstrous animals do not give up without a fight, gripping the whales' skin with their sucker-bearing tentacles; the circular scars on the heads of many sperm whales are the result of dramatic deep-sea battles.

Sperm whales are superb divers—no other mammals can match their ability. Bull sperm whales dive to depths of 6500 ft. (possibly even to 10,000 ft.) in search of food, and judging by the variety of objects (such as pebbles and tin cans) that have been found in their stomachs, they regularly feed off the seabed. Sunlight does not penetrate to such depths, so eyesight is of little use to hunting sperm whales. Instead, the whales rely on echolocation or sonar (the emission and reflection of sound waves) to locate their prey and surroundings. They also possess chemoreceptors in their mouths that can detect the slightest changes in the chemical composition of the surrounding water. The sperm whale has the largest brain of any animal, probably needed to sort out the complicated sonar and "taste" information it receives. The record weight for the brain of a captured sperm whale is 20.3 lbs., compared with a record of 15.2 lbs. for a blue whale.

of thousands of hunted whales, it was discovered that rorquals living north of the Equator have a much more varied diet than those in the Southern Hemisphere. In the cold waters of the Antarctic region, the tiny shrimp-like crustaceans called krill are found in such tremendous concentrations that they dominate the diet of most whales. Since most of the creatures eaten by rorquals do not live far below the surface, the whales do not have to dive very deep for food. However, if danger threatens, they dive to great depths and may not surface for more than half an hour.

Little is known of the mating rituals of rorquals, but it is known that humpbacks mate in a vertical position, at right angle to the surface of the ocean. In most species gestation lasts about a year, and births may take place in any month, although mating activity reaches its peak during the winter.

WHALES AND DOLPHINS—CETACEANS

RIGHT **The Amazon dolphin or boutu lives in the Amazon and Orinoco river systems in South America. Populations living in Bolivia and Peru are located more than 1700 miles up-river from the sea. Amazon dolphins tend to live alone or in pairs, feeding on fish near the riverbed.**

A wax-filled head

A unique feature of the sperm whale—the so-called spermaceti organ—may be linked to the animal's diving and sonar abilities. The huge organ fills the front of the animal's head, and consists of a network of sinuses and nasal passages containing layers of a waxy substance called spermaceti. Some zoologists have suggested that it (and the "melon" organ of dolphins) may work as an "acoustic lens," focusing echolocation signals to improve the efficiency of the sonar system.

Another theory suggests that the spermaceti organ may serve as a buoyancy tank, used when the animal is diving. The flow of water through the sinuses lowers the temperature of the wax, causing it to solidify and become more dense. With a denser body, the whale can descend more easily. As the animal ascends again, the wax melts from an increased flow of blood to the head. This makes the body more buoyant, and helps the whale rise quickly to the surface.

Vulnerable giants

Sperm whales are easier to hunt than rorquals since they have to rest for long periods on the surface of the ocean to recover from long, deep dives. They usually occur in groups, and they tend to encircle and nurse injured animals—a habit that enabled whalers to harpoon great numbers in a small area.

Two special products from sperm whales contributed to the great hunting pressure on the animals—sperm oil produced by the spermaceti gland, and ambergris deposited in the lower intestine. Ambergris is a waxy, gray or black substance thought to be produced by a diet consisting mainly of squid. It is used to make perfume since it smells of musk. The demand for sperm oil declined in the first half of the 20th century, but in the early 1960s its value rose, largely because the US government was using it for its space rocket program. It is ironic that the same space technology that sent a recording of the humpback whale's song into deep space (along with other "messages" from Earth) led to increased hunting of the sperm whale.

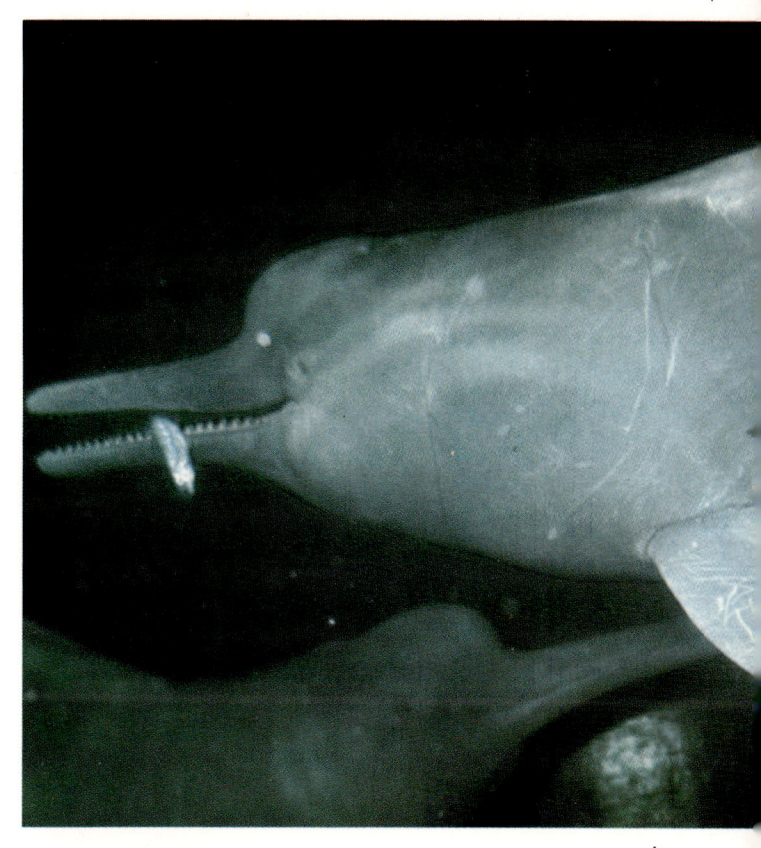

Beaked whales

The 18 species of beaked whales are medium-sized whales, varying in size from over 10 ft. to nearly 45 ft. in length. As their name suggests, they possess prominent elongated jaws, reminiscent of birds' beaks. They live in all the world's oceans, but tend to stay in the open sea beyond the shallow waters of the continental shelf. Since they keep to deeper water, generally avoid ships, and rarely become stranded, scientists have had little opportunity to study most of the species. Much of the knowledge we have comes from dead animals that have been washed ashore, although one species, the northern bottlenose whale, was extensively hunted around the turn of the century.

With the exception of Shepherd's beaked whale, which has well-developed teeth in both jaws, beaked whales have only one or two pairs of teeth confined to their lower jaw. They feed mostly on squid and octopus, both of which can be caught and eaten without the need for a full set of sharp teeth. Males use their teeth in fights, however, and older males tend to have patterns of scars inflicted by rival animals during the mating season. Beaked whales normally live alone or in small groups, though they sometimes congregate in large groups containing up to 40 animals, especially where there is a particularly rich supply of food.

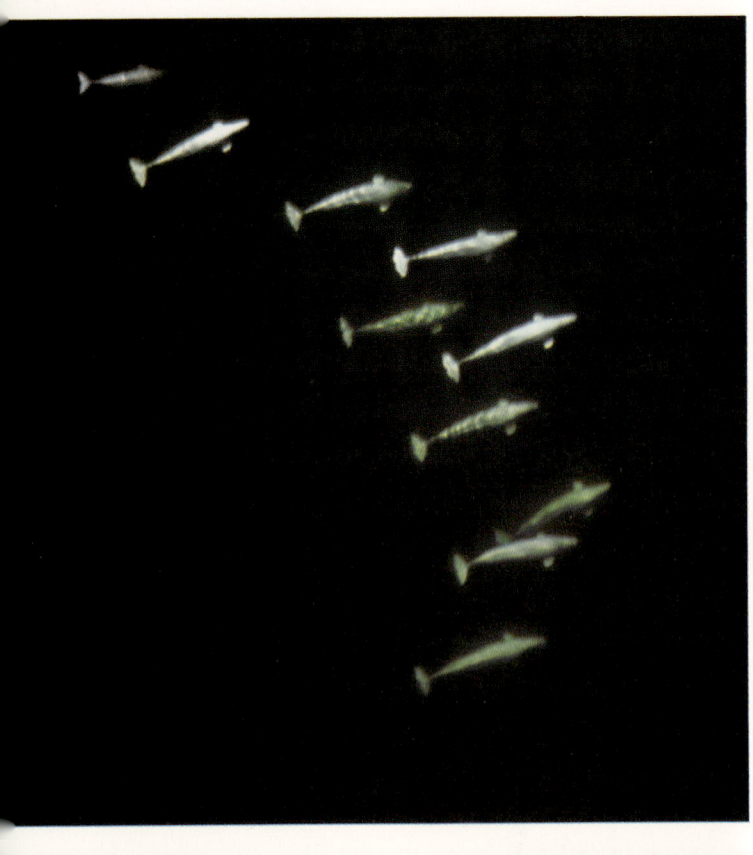

WHALES AND DOLPHINS—CETACEANS

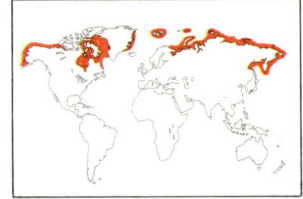

The beluga and the narwhal share a similar distribution—the coastal waters of the far north.

The northern bottlenose whale or barrelhead is the largest of the beaked whales. It grows to 40 ft. in length and has a massive, bulging forehead. It can stay underwater for well over one hour and, like the sperm whale, it has to remain on the water's surface for a long time to recover from deep dives. Along with its greater size and its unusual habit of approaching ships, the northern bottlenose whale's prolonged rests have made the species especially vulnerable to whalers. Though the pressure from hunting has receded, the population is nevertheless thought to be severely depleted.

River dolphins

The five species of river dolphin live in rivers, estuaries and coastal habitats in Asia and South America. They have long, slender jaws full of pointed teeth for catching fish. They also feed on shrimps and, in the case of the La Plata dolphin, squid and octopus. Their eyesight is so poor that it is mainly used to distinguish night from day. Instead of vision, the animals use a highly developed echolocation system to locate and catch their prey. They emit high-pitched clicks and, from the way these sounds are reflected back, they judge the distance of prey and surroundings.

Many river dolphins migrate upstream during the rainy season when rivers are prone to flooding. Unfortunately, in some rivers they now encounter man-made dams that prevent free movement of the dolphins and their prey. Pollution from sewage and industry is another danger to the animals, and the large quantities of silt washed down from deforested hills are increasingly choking many rivers.

The Ganges dolphin lives in the Ganges, the Brahmaputra, Karnaphuli and Meghna rivers in India, Nepal and Bangladesh. It is very similar to the Indus dolphin that lives in the Indus River in Pakistan. Dam construction on the Indus has made this species one of the most vulnerable of all whales and dolphins. The whitefin dolphin lives in the Yangtze River system in China. It is bluish on top and white underneath and its blowhole is on the left-hand side of the head rather than in the center. The Amazon dolphin lives in the Amazon and Orinoco rivers in South America, and is easy to recognize by its pink underside. The La Plata dolphin is the only saltwater dweller in the family. It keeps to the coastline of Brazil, Argentina and Uruguay, and never ventures into freshwater.

TOP Beluga are easy to recognize because the adult's skin is white. They live in northern temperate and Arctic waters, swimming close to the coastline and even venturing up river estuaries in winter. They are gregarious animals that gather in the hundreds during the mating season.
ABOVE Beluga produce an impressive array of sounds, from clicks and whistles to bell-like clangs. Their vocal skills earned them one of their early names—"sea canary."

WHALES AND DOLPHINS—CETACEANS

ABOVE A mature beluga has a pronounced, bulging forehead or "melon" that only reaches full size when the whale is five to eight years old. The animal's beak also develops with age.

White whales

The family of whales known as the white whales comprises just two species—the beluga and the narwhal. Both species have adapted to life in the cold waters of the far north, living off the coasts of Alaska, Canada, Greenland and the USSR.

The adults of the two species are easy to distinguish from other whales. Though the beluga is born a gray color, it passes through a spotted phase to become almost pure white as an adult. It grows to over 16 feet in length, and reaches 1.7 tons in weight. The body of an adult narwhal is a mottled gray-green above and pale beneath, but the most striking feature of the males is the long, spiraled tusk, up to 10 ft. long, extending from the animal's head. Narwhals can grow to 20 ft. in length (excluding the tusk) and reach 2 tons in weight.

Beluga are gregarious animals that usually stay close to the shore. When there is an abundance of food, such as shoaling fish, mollusks, crustaceans and worms, schools may gather containing hundreds of animals. They are cooperative hunters, working together to herd shoals of fish into shallow water. Beluga have unusually flexible necks and can sweep the sea floor with their heads to dislodge stones and flush out prey. They store much of the food as fat to insulate their bodies against the cold Arctic winter. In the summer, beluga migrate together to shallow waters and river estuaries, where females give birth to their calves. Mothers with newborn calves stay away from the rest of the group at first, the infant and its parent swimming together at all times. Calves are suckled for about two years.

Unicorn tusks

Unlike beluga, narwhals prefer the deep offshore waters of the Arctic region. They are not as gregarious as beluga, but females and their young form small groups and males of similar size and age band

587

WHALES AND DOLPHINS—CETACEANS

together. The extraordinary tusk, reminiscent of the horn from the mythical unicorn, is really an extension of one of the narwhal's two remaining teeth. It appears to play a role in ritual displays between males during the mating season.

Narwhals are so well adapted to life in cold water that they can live in seas covered with pack ice. They can break through a layer of ice two inches thick when they need to breathe. Narwhals have long been hunted by Eskimos, who wait at their breathing holes ready to spear the surfacing whales with hand-held harpoons. Narwhals have been commercially hunted for their meat, skins, oils and tusks, and the population, estimated at less than 20,000 in 1981, is steadily declining.

The porpoises

The porpoises are the smallest members of the whales and dolphins group, some adults reaching no more than 5 ft. in length. The six species are a combination of black, gray and white in color, and most have a small triangular dorsal fin. They generally live in coastal waters, though the largest species, Dall's porpoise, ranges into the deep waters of the North Pacific. Porpoises have varied diets, including prawns, squid, and fish up to the size of cod. They generally live in small groups—large concentrations are usually temporary gatherings at good feeding grounds.

The harbor or common porpoise is the most widespread species, occurring in both the North Pacific and the North Atlantic, as well as the Bering Sea, the Baltic Sea and the Black Sea. Dark on the upperside and pale on the underside, it often approaches close enough to the shoreline to be seen from dry land.

ABOVE LEFT AND RIGHT Like many dolphins, the tucuxi of South America frequently leaps from the water. Though it has been suggested that dolphins sometimes leap for pleasure or that such "breaching" may have a social role, it also clearly helps them to catch prey. The impact of the dolphin hitting the surface of the water frightens fish, so that the shoals concentrate into tight bunches. Once the fish have amassed, it is easy for the dolphins to catch them and feed.

The dolphin family

The dolphin family is large and varied, containing 32 species that live in all the oceans of the world. The largest and most imposing of them is the killer whale, or orca. The best loved is probably the bottle-nosed dolphin, the star of many films and oceanaria. Other species include the common dolphin, the spinner dolphin, Risso's dolphin, the tucuxi, the melon-headed whale, the false killer whale and the pilot whales. Dolphins have been hunted mercilessly since prehistoric times, and thousands are killed accidentally every year, drowned in fishing nets, caught up in the herding operations of tuna fishermen, or killed by pesticide pollution.

Dolphins are, with the exception of killer whales, small to medium-size animals measuring 4-13 ft. head to tail, streamlined with a backward-curving dorsal fin, a single blowhole, a pronounced beak and a set of sharp teeth, well separated for catching fish. Dolphins with well-developed beaks are generally fish eaters; those with a less pronounced beak and with less teeth feed mainly on squid. The bulging forehead in most species may be related to the development of the brain's complex echolocation system which is used to find fast-moving prey.

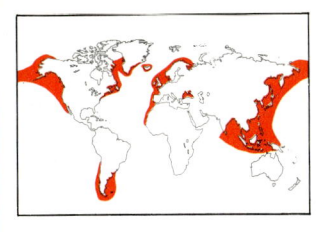

WHALES AND DOLPHINS—CETACEANS

Most porpoises live in the Northern Hemisphere; some keep to coastal waters, but others occur in the open ocean.

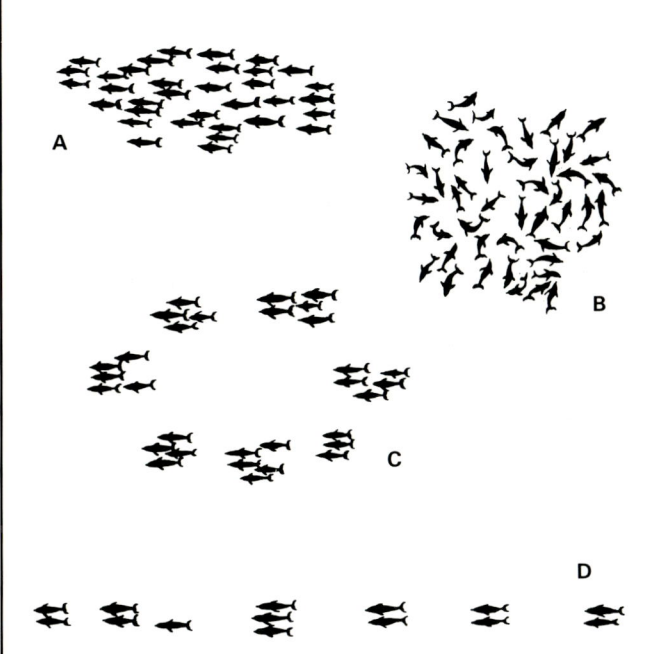

ABOVE Common dolphins are one of the most numerous of all the whales and dolphins, and may occur in schools numbering many thousands.
ABOVE LEFT A school of common dolphins normally swims in a loose formation (A), the pattern breaking into swirling confusion as the school comes across a shoal of fish (B). The oval formation of bottle-nosed dolphins is an ordered, ritual swimming pattern (C), like the occasional straight-line formation adopted by common dolphins (D).
PAGES 590-591 Bottle-nosed dolphins can reach speeds of over 20 mph, regularly leaping right out of the water as they race along.

Color and feeding

The color of dolphins varies greatly and depends on the areas where they feed. Dolphins with large splashes of color generally feed near the surface of the water, where the difference in shades and the patterns of sunlight blur the outline of the animal. This effectively camouflages them from predators such as sharks, as well as from the dolphin's prey. Plain-colored dolphins feed mostly in the deep sea where their grayish coloration blends in with their dim surroundings.

A liking for company

Dolphins are gregarious creatures. While coastal dolphins are found in groups of two to a dozen individuals, those species that feed out at sea form much larger groups, containing as many as 1000 animals. They often work together to herd shoaling fish, such as anchovy. At certain feeding sites, as many as 2000 dolphins may gather. Most dolphin species do not have a stable social structure; they do form groups, but individuals leave and join them constantly. Risso's dolphin and the killer whale, however, form stable groups comprising a mature male with a harem of females and their young.

In the search for food, some species make migrations from shore to deep sea, or from deep sea to shore. Others make north-south migrations every year. Dolphins are opportunistic feeders, taking food when and where they can. Since ancient times they have been known to follow ships sailing far out at sea, probably in the hope of finding food scraps.

Battling dolphins

Sexual activity peaks during the summer months when bloody fights take place between males who chase each other at great speeds. When they clash, one dolphin may bite or run its teeth over an opponent's back. Sometimes a male will point his head at a

589

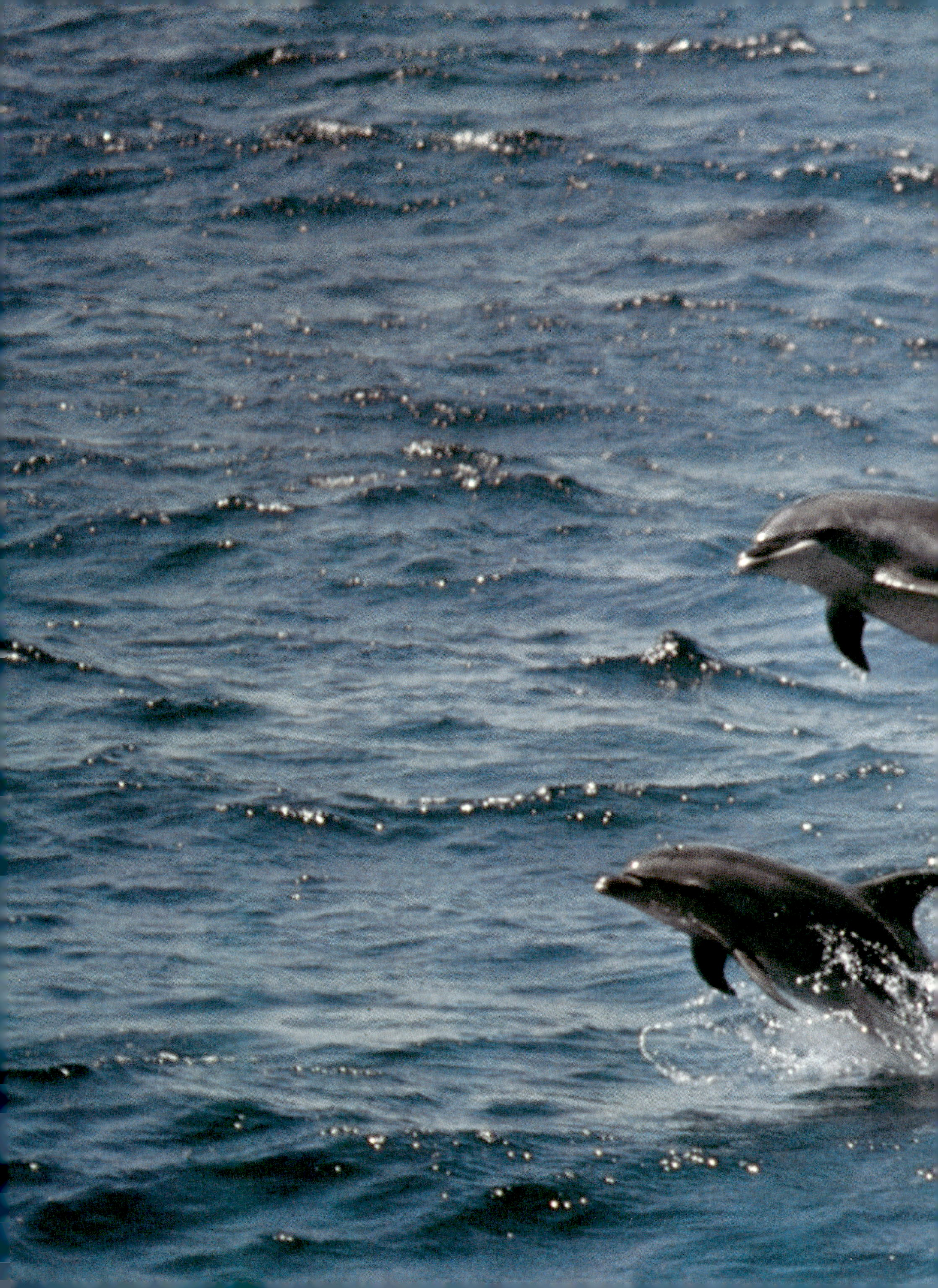

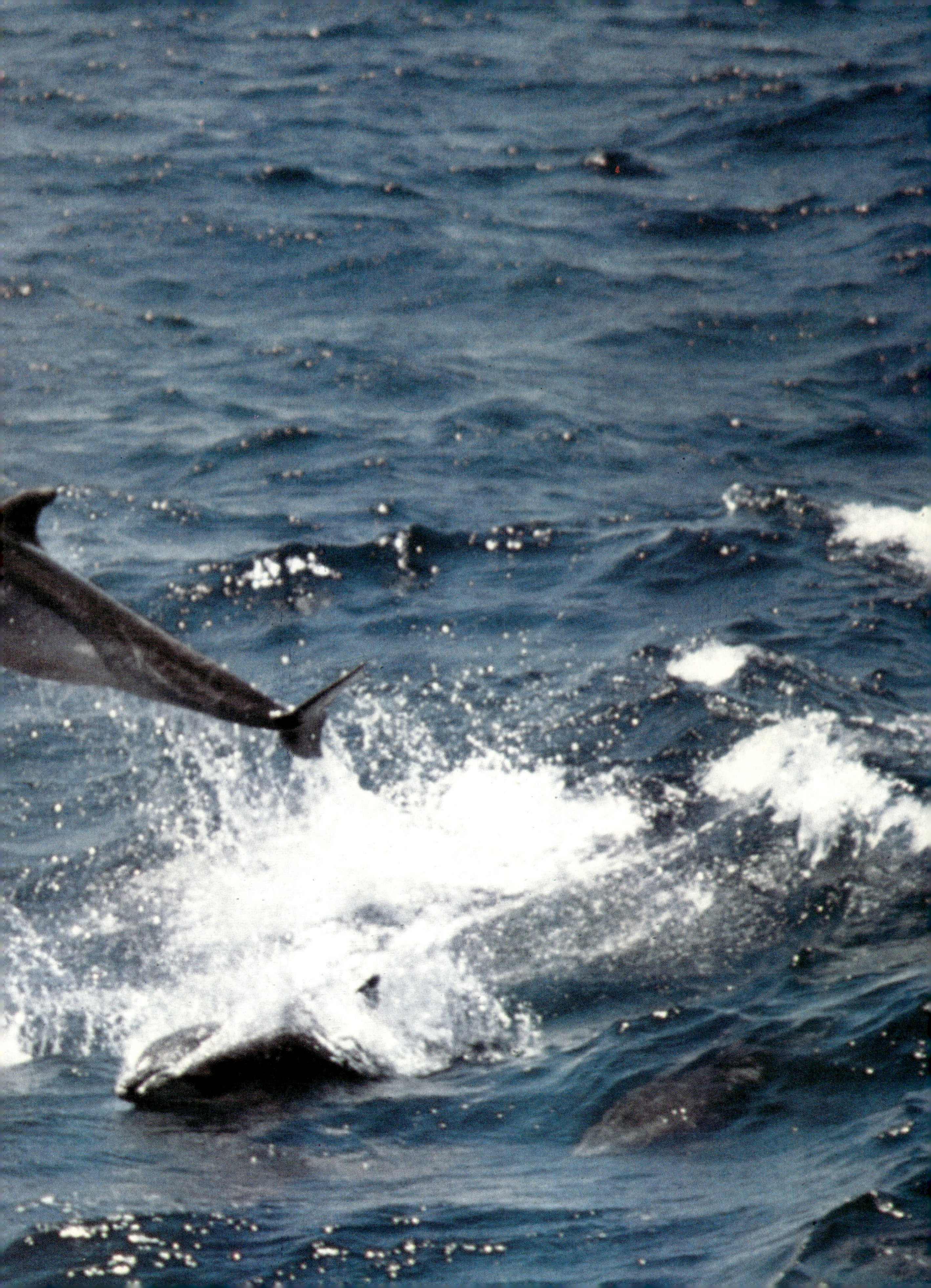

WHALES AND DOLPHINS—CETACEANS

LEFT The Pacific white-sided dolphin lives in the temperate waters of the north Pacific Ocean. It has a short, black beak and a rounded snout.
BELOW LEFT Commerson's dolphin occurs in the cold seas around southern South America and the Falkland Islands. Its striking piebald coloration makes it one of the easiest dolphins to recognize.
BOTTOM LEFT Of all species of whales and dolphins, pilot whales are the most likely to become stranded. Whole herds are sometimes washed up on a beach and, unless they are helped back into water that is sufficiently deep for them to escape, they quickly perish.

potential rival and open and close his jaw quickly as a threat. Behavior of this sort between dominant males and challengers has been observed among dolphins held in captivity, but fights between dolphins do not usually lead to fatalities.

When a female dolphin finally allows the male to approach her, she rolls over onto her back on the water surface and the male lies on top of her, both disappearing below the surface for a while. Pregnancy lasts 10 months, after which one offspring is born. Twin births do happen but are rare. Since sharks are attracted by the smell of the blood shed during childbirth, it is a dangerous time for dolphins and their young. Mothers are very attentive to their calves, but there seems to be no bond between fathers and their offspring.

Most dolphins are fast swimmers. Their speed enables them to go after fast-swimming fish and to escape sharks. Many dolphins jump out of the water in apparent playfulness, but one purpose of jumping may in fact be to frighten the shoals of fish that they chase, especially when a group is hunting together.

Mass beaching

Schools of certain dolphins—notably pilot whales, melon-headed whales and killer whales—are regularly found stranded on beaches. In some cases, more than 100 animals have been found dying on a beach at the same time. Why these large dolphins (along with sperm whales and pygmy sperm whales) should perish in this way is not clear. It may be that outsized waves carry them too far toward the shore when they are chasing shoals of fish in the shallows. Other possible causes include disruption of the whales' navigation systems, and disturbances to their sense of balance by parasitic worms that infect the inner ear.

WHALES AND DOLPHINS—CETACEANS

TOOTHED WHALES CLASSIFICATION

The toothed whales form the suborder Odontoceti—part of the order Cetacea. They are far more numerous than the baleen whales, comprising 66 species in all, grouped into six different families.

The family Physeteridae contains three species, all ranging across the temperate and tropical oceans of the world: the sperm whale or cachalot, *Physeter macrocephalus*, and the much smaller pygmy sperm whale, *Kogia breviceps*, and dwarf sperm whale, *K. simus*. The 18 species of beaked whales form the family Ziphiidae, found in the deep sea throughout the world. They include Cuvier's beaked whale, *Ziphius cavirostris*, the northern bottlenose whale, *Hyperoodon ampullatus*, and Sowerby's beaked whale, *Mesoplodon bidens*.

The family Platanistidae, the river dolphins, contains four species that inhabit major river systems, including the Ganges dolphin or Ganges susu, *Platanista gangetica*, and the Amazon dolphin or boutu, *Inia geoffrensis*. A fifth species, the La Plata dolphin or franciscana, *Pontoporia blainvillei*, lives along the coast of Brazil, Uruguay and Argentina in South America.

Only two species belong to the family Monodontidae, the white whales: the beluga or white whale, *Delphinapterus leucas*, and the narwhal, *Monodon monoceros*, both of which occur in the cold waters around the northern USSR, Alaska, Canada and Greenland. There are six species of porpoises in the family Phocoenidae. They include the harbor or common porpoise, *Phocoena phocoena*, which inhabits northern coastal waters; Dall's porpoise, *Phocoenoides dalli* of the North Pacific; and the finless or black porpoise, *Neophocoena phocoenoides*, from parts of the Indian and Pacific oceans.

The dolphins form the largest family, the Delphinidae, with a total of 32 species in 17 genera. Species include the bottle-nosed dolphin, *Tursiops truncatus*, which lives in coastal waters around the world; the common dolphin, *Delphinus delphis*, from most tropical and warm temperate seas; the tucuxi or estuarine dolphin, *Sotalia fluviatilis*, of the Amazon river system and the eastern coast of South America; the false killer whale, *Pseudorca crassidens*, of the warmer oceans; and the killer whale or orca *Orcinus orca*, which occurs in oceans throughout the world.

RIGHT A group of common dolphins helps an injured companion to keep afloat. Such behavior is typical of many dolphins, but it has proved disastrous for some animals—hunters have been able to harpoon large numbers as the dolphins rally around to help one another.
BELOW The sequence of drawings shows how a common dolphin makes a spinning leap from the water.

THE BOTTLE-NOSED DOLPHIN
— MASTERY OF THE WAVES —

The bottle-nosed dolphin is the dolphin most familiar to countless visitors to oceanariums and aquariums. Its ability to leap more than 20 feet out of the water to pluck fish from a trainer's hand never fails to astonish the crowds. Found in the coastal waters of most temperate, tropical and subtropical oceans, the bottle-nosed dolphin is larger than the common dolphin, growing to around 11 ft. long. But it is less colorful, being dark gray on top, paler on the sides and with a white-pink coloring underneath. It has a pronounced beak with 40 teeth in each jaw, a curved, sickle-shaped dorsal fin (on the back) and curved pectoral fins (on the side).

Like all dolphins, the bottle-nosed is a fast swimmer: an adult can reach 22 miles an hour. It is a sociable animal, forming schools of several dozen individuals of both sexes and all ages, apparently without an obvious leader. They usually stay close to the coast in a tightly bunched group and may sometimes venture up river estuaries. When one of the group is injured, two others will position themselves on either side of it, shove their noses beneath its flippers and push the animal to the surface so that it can breathe. Like killer whales, the dolphins work together to herd shoals of fish. Their eyesight is poor and their sense of smell almost nonexistent. They rely instead on echolocation to find the fish, cuttlefish (a kind of mollusk) and shrimps on which they feed.

A language of whistles and squawks

Bottle-nosed dolphins use a basic language made up of a vast array of sounds to communicate with one another. These include whistles, clicks, squawks and grunts. By means of these calls, mates can recognize one another and a calf can recognize its mother. It is believed that bottle-nosed dolphins can even identify the sound of certain ships. If a vessel has hunted them in the past, they will stay well out of reach until danger has passed.

An after-dinner nap

The dolphin dives to depths of around 65 ft. in order to catch fish. When underwater, the animal's pulse rate drops from 110 beats per minute to just 55. With the blood pressure reduced, the blood takes oxygen mainly to the vital organs—the heart and brain—allowing the dolphin to stay for a longer time underwater. After a good feed, the dolphins sleep for a short time. Females lie asleep on the water surface with their blowholes exposed; males lie just beneath the water, coming up for breath every now and then through a reflex response to lack of oxygen.

Nursing companions

Females reach sexual maturity at the age of five years; males mature about three years later. Breeding lasts from spring to summer. After a gestation period of one year, the pregnant female leaves the school with two female companions that protect her from shark attacks and nurse her when she goes into labor. One offspring is born at a time. The calf, born tail-first, is brought immediately to the surface to

594

take its first breath. It can swim rapidly soon after birth and keeps close to its mother's side. After two weeks it begins to swim on its own, sometimes pursuing fish, but rarely catching them, and it will rush back to its mother at the first sign of danger. The calf's teeth are not suitable for chewing solid food until the animal is several months old, and it continues to be suckled for up to a year. During this difficult period the mother always has a female helper by her side, for a young dolphin is easy prey for a large shark and a mother cannot feed, teach and be lookout all at once.

RIGHT A bottle-nosed dolphin uses its powerful tail to propel itself backward.
BELOW The streamlined shape of the dolphin, with its curved dorsal fin, makes it a fast and agile swimmer.
BELOW LEFT A pair of bottle-nosed dolphins await the reward of a fish after performing for their trainer.

WHALES AND DOLPHINS—CETACEANS

Killer whales

The killer whale is the largest member of the dolphin family. Males grow up to 35 ft. long and are larger than females. They are easy to recognize, with their striking black back, white belly and white patch behind each eye, and by their large dorsal fin that stands up to 6 ft. high.

A pod of whales

Killer whales live in all oceans, but they are more common in Arctic and Antarctic waters where there is a plentiful food supply. They are social mammals, living in groups (or pods) of as many as 40 individuals. A typical pod contains one dominant male, four adult females and several immature whales of both sexes. They eat mainly fish and squid but they also prey on dolphins, porpoises, seals and sea birds. (The killer whale is the only member of the whale and dolphin order to prey on warm-blooded animals.)

Killers that cooperate

The whales are cooperative hunters, both adults and immature animals working together to herd shoals of fish into a particular area. Groups of killer whales even combine to attack the great baleen whales, tearing into their victim's most sensitive spots—the stomach, genital region, and tongue. Yet, despite its fierce reputation, the killer whale has not been known to attack man.

Killer whales will remain in the same area all year round if plenty of food is available there. Others travel up to 300 miles in search of food. Sometimes more than one pod may gather on rich feeding grounds close to the coast, and this may give the impression that the killer whale population is booming. In fact, in many areas they have been killed in large numbers to protect local fisheries.

Females reach sexual maturity at eight years old. Males mature much later, at 12-14 years. Gestation lasts for 15 months, after which one calf is born. Calves are suckled for a minimum of a year, but they stay close to their mothers for at least three years. Females spend the rest of their lives in the pod they were born in, while males may go off in search of another one. The establishment of hierarchical positions, and the splitting up and regrouping of pods that have grown too large, seem to be resolved without a great deal of aggression.

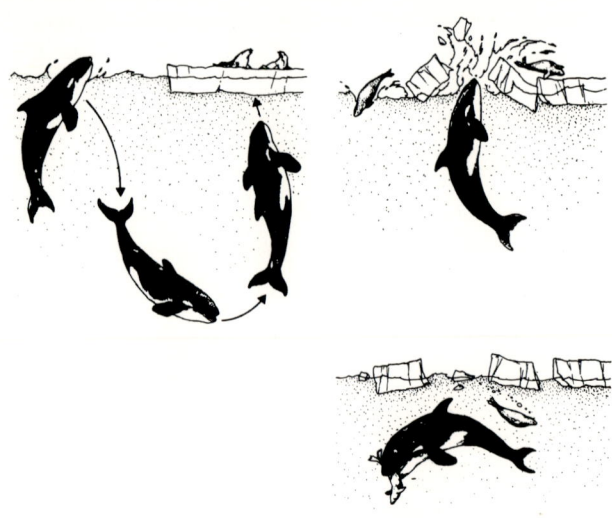

TOP The killer whale is the only member of the whales and dolphins group to prey on warm-blooded animals. Its broad diet includes seals, penguins, dolphins, porpoises and even slow-swimming whales. Since fish and squid form the major part of their diet, killer whales sometimes come into competition with fishermen and have been hunted in the name of "fishery protection."
ABOVE Killer whales are known to break through thick layers of ice to force resting seals into the water.

PLATYPUS AND ECHIDNAS—MONOTREMES

EGG-LAYING ODDITIES

Alone among the mammals, the echidnas and the platypus do not give birth to live young—instead, the females produce soft-shelled eggs from which the offspring hatch

PLATYPUS AND ECHIDNAS—MONOTREMES

The monotremes—echidnas and the platypus—have probably excited more controversy than any other group of mammals. Indeed, when the first reports of these strange creatures reached the zoologists of Europe from Australia, many doubted whether they were to be trusted. The dried platypus skin that appeared in Britain in about 1798 was thought to be a fake, assembled by a taxidermist from the body of a mammal and the bill and feet of a duck. But on further investigation, no tell-tale stitching could be found, and the arrival of complete specimens proved that the animal was genuine.

Examination of the reproductive system suggested that the animal laid eggs, like reptiles. But since it also appeared to suckle its young, the egg-laying theory was rejected. When subsequent evidence proved that the platypus did indeed lay eggs, zoologists were left with a problem—how to classify it. At first, they thought of placing it among the reptiles: platypuses were referred to as "furred reptiles." Certain skeletal peculiarities of the platypus, similar to the bone structure of lizards, reinforced this idea.

Following this theory, the platypuses were assumed to be at least partially cold-blooded (that is, unable to regulate their own body temperatures accurately, and at the mercy of outside temperatures). This was disproved only recently, when careful studies of the platypus showed that it was fully warm-blooded and active throughout the year. Today it is recognized that, despite their bizarre appearance and unusual habit of laying eggs, the monotremes are basically as mammalian as we are.

Spiny anteaters

The echidnas, also known as spiny anteaters, have many features in common with the pangolins and giant anteater. They have extended, probing snouts with small mouths, long sticky tongues and strong claws. Their bodies are covered with strong spines distributed in patches among the hair on their backs. Like European hedgehogs, they curl up to conceal their vulnerable underparts within a protective spiny ball.

Skeletal evidence

Although monotremes are fully developed mammals with typical mammalian skeletons, they retain many features in common with their reptilian ancestors. The structure of the skull, for example, resembles that of

TOP Sleek and glistening from the river, a platypus probes the mud of a backwater for shellfish, insect larvae and worms.
ABOVE The platypus bill is extremely sensitive. Recent research has shown that it contains electroreceptors able to detect the smallest changes in the electrical field around the animal—even that caused by the flick of a shrimp's tail.
PAGE 597 A furry mammal with a bill and webbed forefeet similar to a duck's, and a flattened tail, the platypus looks so odd that it was initially thought to be a fraud put together by taxidermists.

PLATYPUS AND ECHIDNAS—MONOTREMES

reptiles. Whereas the bones of the cranium knit together with clearly visible joins in most mammals, those of monotremes are fused with no trace of a joint; this indicates that the skull becomes fully formed at a very early stage in the animal's life—much earlier than is usual for mammals.

Adult monotremes have no teeth; their function has been taken over (in the platypus) by strong bony plates that are used to grind prey, and (in the echidnas) by two sets of spines—one at the base of the tongue, another on the roof of the mouth—used to crush ants and termites.

The poison spur

Male monotremes are venomous mammals, a characteristic they share with certain shrews. Echidnas have the equipment to make poison, but they no longer use it. Male platypuses, however, are still capable of poisoning predators.

The platypus manufactures its poison in a gland in the thigh, just behind each hind knee, and delivers it to a hollow spur (about 0.8 in. long) on each heel. The spur is normally covered by a fold of skin, but it can be protruded at will and stabbed into an enemy using a strong backward jab. The poison is a transparent liquid that kills victims by causing their blood to thicken. A male platypus is quite capable of killing a dog this way, and although the poison cannot kill a full-grown man, it causes severe pain.

The venom glands grow larger during the breeding season, and the poison becomes more toxic, suggesting that the purpose of the poison is connected with mating. Males will certainly fight using their spurs in an attempt to retain control of their breeding territories. However, a lethal poison is not an ideal way to cope with rivals of the same species, and it is more likely that the spurs were developed as a form of defense against a predator that is now extinct.

An all-purpose outlet

Like birds and reptiles, the echidnas and platypus have two uteri; placental mammals have only one. Monotremes do not have a separate birth canal through which they lay their eggs. The eggs pass through the same chamber (the cloaca) that the animals use for excretion of feces and urine—hence the name of the order "Monotremata" which means "one-holed creature."

ABOVE A short-nosed echidna makes a midday foray across the scorched earth, sniffing the ground for traces of ants and termites. Like many other Australian mammals, it has evolved features that recall those of other, quite unrelated mammals on distant continents—in this case the narrow snout, long, sticky tongue and powerful claws of the anteater or the pangolin. Since it exploits the same food source, the echidna has evolved the same effective tools of the trade.

The eggs are rich in yolk and are soft-shelled like reptile eggs. Platypus eggs are sticky when laid. They hatch after an incubation period of about 10 days. As with reptiles and birds, the young monotremes emerge from the egg by breaking the shell with an "eggtooth" that grows in front of the jawbone and disappears as the young animal grows. At the time of hatching, the animals are extremely small: a young echidna measures only about 0.6-0.8 in. long, and a young platypus about 1 in.

The birth of an echidna

Echidnas lay only one egg at a time. It emerges from the cloaca as the female lies on her back, and is deposited in her "incubatorium." This is a deep pocket on her belly, formed of enlarged folds of skin and

599

PLATYPUS AND ECHIDNAS — MONOTREMES

LEFT **Awkward on land, the platypus is in its element in water, propelling itself with its powerful, webbed front feet and steering with its hind feet and flattened tail.**
BELOW LEFT **Platypuses have two kinds of burrows. The burrow on the right, opening onto the riverbank, is used as a general shelter by both sexes. The one on the left is a special breeding burrow where perhaps the most remarkable aspect of this strange mammal's life takes place. In a nesting chamber lined with leaves, the female gives birth to two eggs. Soft-shelled like those of a snake or lizard, and similar in size to a house sparrow's egg, they hatch in about ten days. The young feed on their mother's milk for another four or five months before leaving the nest, by now fully furred.**

muscle, and resembling the pouch of a marsupial. It differs in being temporary, for it develops only after the egg has been fertilized. The female platypus does not develop a pouch on the stomach. Instead, she digs a complex nesting burrow in the ground that may extend for as much as 100 ft. and have several underground chambers.

Milk patches

As early zoologists observed, young monotremes feed on their mother's milk. Unlike other mammals, however, they do not obtain the milk from well-defined teats. They suck it from patches of milk-soaked fur surrounding simple apertures where the mammary glands open onto the surface of the skin. The female platypus has two mammary glands on her underside.

In the echidnas, the mammary glands emerge inside the stomach pouch, enabling the newly hatched animal to suckle without leaving its snug shelter.

Echidnas and platypuses suckle in different ways. Young echidnas suck vigorously at the milk patch, and this in turn stimulates the secretion of milk. A young platypus, on the other hand, merely licks up the liquid that trickles into its mother's fur. When it wants to drink, the baby climbs onto its mother's belly and lightly taps the mammary gland with its bill. The mother platypus, lying on her back, contracts her ventral (stomach) muscles to force the milk out and the young then drinks the milk that emerges.

A peaceful evolution

The monotremes of today have frequently been referred to as a link between reptiles and other mammals, but their precise place in the evolutionary story is hard to define. They are probably the survivors of a group of early mammals that started to evolve separately some 180 million years ago. This is well before the marsupials emerged as a separate group, and certainly long before the development of major groups such as the carnivores, ungulates, rodents and primates. These early mammals migrated into the Australian continent before it broke away from the other continents. They were able to develop without suffering competition from the more efficient placental mammals, which evolved after the rift opened up between Australia and South America. Isolated from competitors and enemies by the movements of the Earth's crust, they became highly adapted for their environment, and remarkably successful.

PLATYPUS AND ECHIDNAS—MONOTREMES

EGG-LAYING ODDITIES

The solitary echidnas

There are two species of echidna: the short-nosed echidna of eastern mainland Australia and many of the nearby islands, Tasmania and southern and central New Guinea; and the much rarer long-nosed echidna, which is restricted to the humid forests of the New Guinea highlands, and possibly Sulawesi Island, Indonesia, to the west. Both species are spiny, but the long, porcupine-like spines of the short-nosed species are more prominent.

Long-nosed echidnas have long fur that partly conceals their spines. If attacked by a predator, they roll themselves up into an impregnable ball of spines. Their keen senses of hearing and smell, however, often warn them of enemies well in advance, allowing them to take cover. On soft ground, an echidna digs straight down into the soil, sinking below the surface until only a few spines can be seen emerging from the disturbed ground.

Hunting for ants

Echidnas are woodland animals and lead solitary nocturnal lives, sniffing among the leaf litter and undergrowth in search of prey. The short-nosed echidna feeds on ants and termites, licking them up with its long, sticky tongue. The echidna locates its prey using its keen sense of smell. When it discovers an ants' nest or termite mound, it demolishes the structure with a few blows of its great, clawed forelimbs and then scoops up the insects as they try to escape. It invariably manages to swallow not only the termites (or ants) but also much nest material, such as

BELOW A pair of long-nosed echidnas defend themselves from danger by lying half buried in the ground, presenting only their spine-covered backs to a potential enemy. Echidnas use their powerful claws to dig furiously into the soil. If the ground is too hard, the echidna may roll up like a hedgehog to hide its soft underparts within a prickly armor of spines. It may even defend itself by spitting half-digested food at its attacker.

PLATYPUS AND ECHIDNAS—MONOTREMES

small pieces of wood, pebbles, soil and sand. All this is put to good use, helping to break down the strong armor of the insects in the echidna's stomach.

Where food is abundant, as in the rain forests, each echidna forages over a regular range of about 125 acres; if food is harder to find, they travel farther afield. They are normally most active at twilight, but alter their habits with the temperature. In winter they often emerge to hunt during the day, taking advantage of the sun's warmth. But in a hot summer they will do the opposite, resting by day and foraging all night.

An earthworm diet

The long-nosed echidna is a more elusive creature, and its habits have yet to be studied in depth. Fossil evidence indicates that similar animals were found throughout Australia until about 10,000 years ago. Today the long-nosed echidna is restricted to wet, tropical mountain forests. The probable reason for this is its diet. Instead of eating ants and termites—readily available throughout the arid regions of Australia—it feeds almost exclusively on earthworms, which live only in the moist, rich soil typical of grasslands and broad-leaved forests.

Despite the different shape of its prey, the long-nosed echidna's mouth (with its long tongue) and its tubular snout are much like those of its ant-eating cousin. To swallow a worm, the echidna gathers it up headfirst with its tongue (equipped with bristles for that purpose), and hauls it into its snout—like a human sucking up a piece of spaghetti.

The mysteries of mating

At present, relatively little is known of the sexual habits of the echidnas. The mating season for the short-nosed species lasts from June to August—midwinter in Australia. During this period, the females attract males by leaving a scent trail as they forage. The males presumably vie for the female's attentions, since a number of them may congregate round her at a time, but the details of sexual rivalry and mating are still uncertain.

The female echidna digs a burrow in which to lay her eggs. When they hatch, young echidnas are hairless and their snouts are short and rubbery. They stay in the mother's pouch for seven or eight weeks until their spines begin to grow. By this time they measure 3 to 3.5 inches long and are independent enough

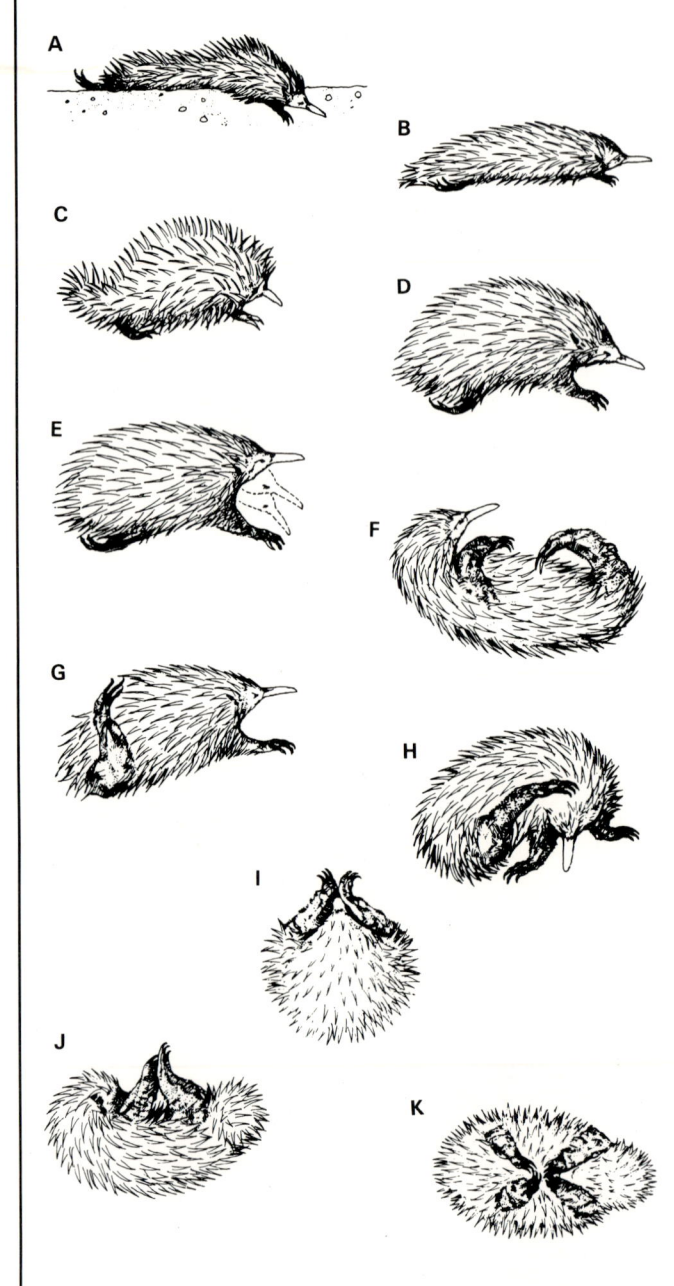

ABOVE The behavior of a short-nosed echidna: sleeping (A); in a submissive position (B); aggressively displaying its spines (C); foraging for food (D); lifting its head to sniff the air (E); scratching its belly, back, and head (F, G and H); and rolled up in self-defense (I, J and K).
FAR RIGHT A long-nosed echidna stands on its hind legs as it searches for earthworms—its favorite food—in a raised part of the forest floor. It uses its elongated hind claws for grooming and digging.

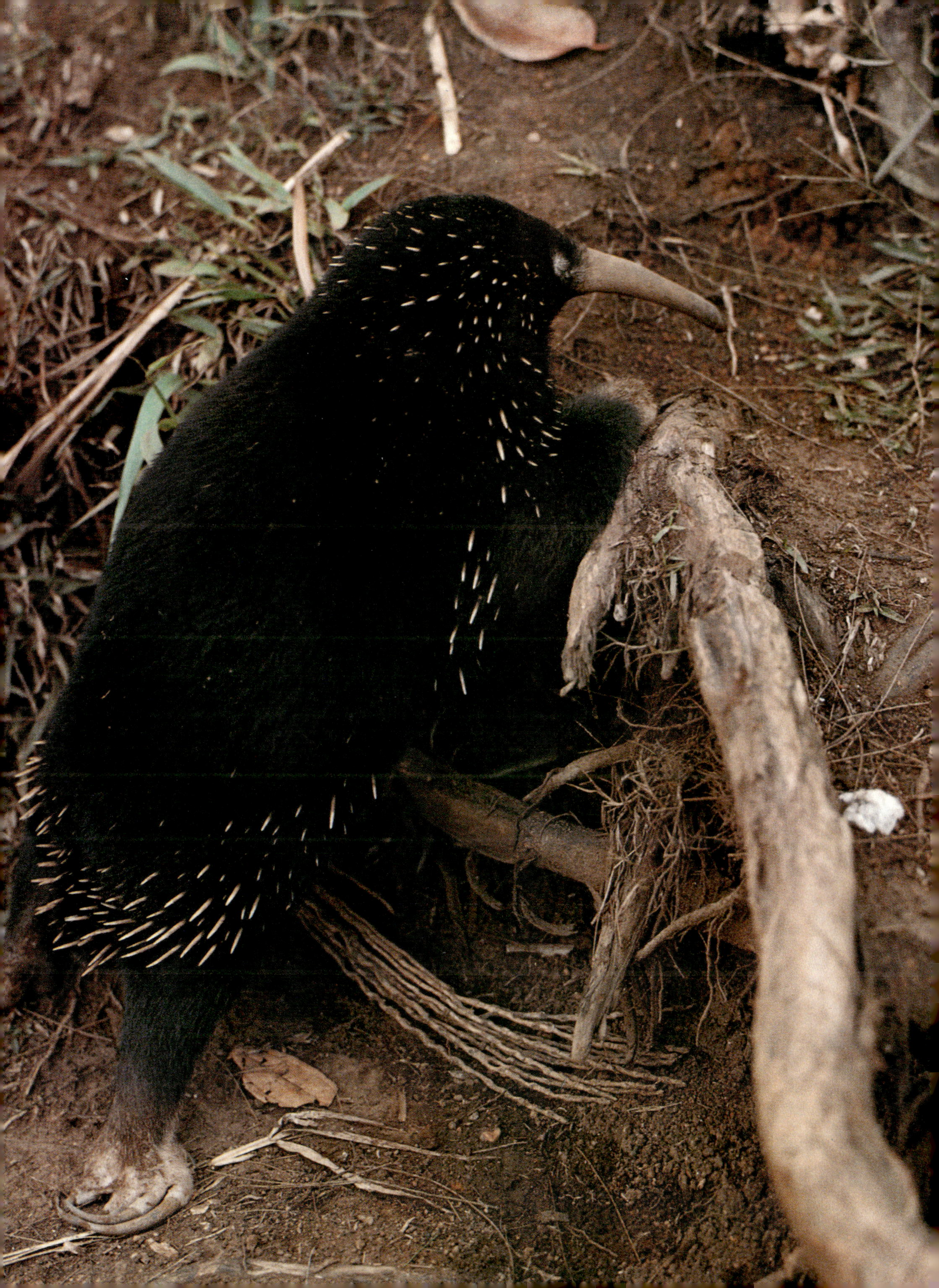

PLATYPUS AND ECHIDNAS—MONOTREMES

ABOVE Echidnas live a surprisingly long time. One short-nosed echidna that was kept in a zoo lived for more than 50 years. Out in the wild their life expectancy is probably shorter, although they have few natural enemies today. Their spiny defenses may have evolved to protect them from predators that are now extinct (such as the marsupial lion, once common in Australia).

to be left by themselves, although their mother continues to suckle them for about six months. Once they are weaned, they begin to explore the area around the nest by sniffing with their long snouts for food. They become sexually mature and completely independent about 12 months after hatching.

Origins

The many unique features of the monotremes have inspired a great deal of speculation about their origins. So far, fossils of early monotremes have only been found in Australia. The earliest platypus-like fossil dates from the mid-Miocene epoch some 10 million years ago. One theory of their origin suggests that monotremes developed in the Americas. South America and Australia both have their own species of marsupial, indicating that the primitive animals shared the landmass before the area that is now Australia broke away. However, so far no fossils have been found in South America to support this theory.

The monotremes became isolated from the mainstream of mammalian evolution, developing characteristics unique to themselves. They never developed the advanced mammalian features such as live-born young, well-formed teats (both found in the relatively primitive marsupials) and feeding of the unborn young via a placenta.

Furthermore, the monotremes' high degree of specialization indicates that, in all likelihood, they broke away from the mainstream evolutionary path a very long time ago. Features such as a duckbill and poison glands are the result of millennia of evolution. They are not a relic of some primitive ancestor. If they were, species from different branches of the evolutionary tree would also possess these features.

The amphibious duckbill

Many of the unusual features of the platypus (such as its bill and webbed feet) are adaptations that equip it perfectly for its chosen habitat: the streams, rivers and ponds of eastern Australia and Tasmania.

The male, only slightly larger than the female, is about 18-24 in. long and weighs 2-5 lbs. 8 oz. while the female measures 16-22 in. long and weighs 1 lb. 8 oz.-3 lbs. 5 oz.

PLATYPUS AND ECHIDNAS—MONOTREMES

RIGHT **Aspects of platypus behavior:** swimming on the surface with its nostrils out of the water (A); swimming to the surface for air after a long dive (B); calling during the breeding season (C); sitting upright on its hind feet using its tail as a support (D); probing riverbed mud in search of food (E); grooming the fur on its belly (F); suckling its young, which lick the milk from the fur around the mammary glands (G); and excavating a burrow with its strong, heavily clawed forepaws (H).

The male has a heavy, thick-set body covered with a thick coat of soft brown fur, interspersed with bristles. Its ears, lacking external ear flaps, are concealed by fur.

A sensitive bill

The most striking feature of the platypus is its bill. It measures about 2.5 inches long in the male and 2 inches in the female. Broad, flattened, and covered with shiny black skin, it is very like the bill of a duck—and in many areas the animal is known as the duckbill or duckbilled platypus. The bill is not just a horny alternative to teeth: it is a sensitive organ, equipped with nerve endings that allow the platypus to probe the bed of the river or lake for food. Its eyes and ears remain closed when it swims underwater.

Another duck-like characteristic of the platypus is its webbed feet. These give it considerable power and maneuverability in the water. Each foot has five toes, with long, sturdy claws that are used for digging. The forefeet do most of the work, whether the platypus is swimming, digging or walking. On land, the animal walks clumsily by dragging itself along with each step.

A tunnel builder

By contrast, the platypus is a good swimmer and an excellent digger. Its tunneling skills are used to construct burrows along the banks of rivers and lakes. These are complex in design, each burrow consisting of a deep, roomy central chamber with several tunnels leading off. Breeding burrows are the most elaborate, sometimes having several nesting chambers. The tunnels, often several yards long, weave in and out of one another to form a maze. Some emerge under the water surface, while others open onto the bank. The complexity of the structure is probably intended to confuse intruders and allow the platypus to escape if attacked, although today it has no natural predators capable of penetrating such a tunnel system.

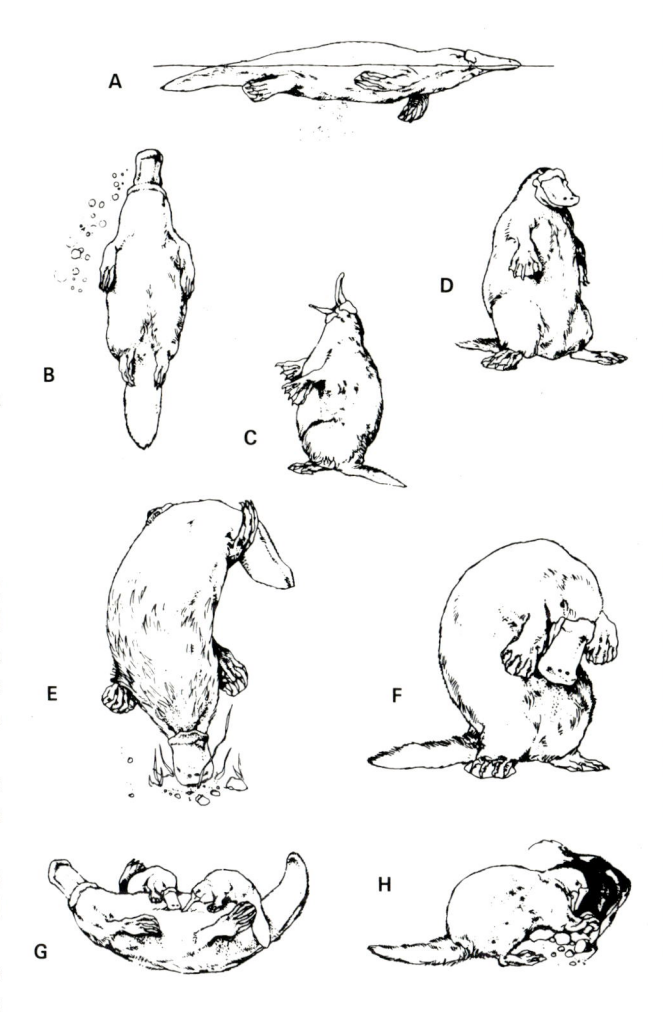

The platypus emerges from its burrow to forage as night falls, slipping into the water and submerging with a swish of its broad, flattened, beaver-like tail. The tail measures about 4-6 in. long in the male and 3-5 in. long in the female. The platypus swims with its bill slightly raised, breathing through the nostrils at its tip. When it tires of swimming, the animal will float idly on the surface of the water.

The platypus feeds almost exclusively on the creatures that live on the river or lake bed: shrimps and other crustaceans, insects, shellfish and worms. It finds these by sifting through the mud and slime on the bottom with its sensitive bill, and filtering the disturbed sediments to pick up anything edible. It does not appear to eat plants of any kind. The cheek-pouches, similar to those of a hamster, contain horny ridges that are used to grind down the tough casings of the shellfish and insects before they are swallowed.

PLATYPUS AND ECHIDNAS—MONOTREMES

The platypus and the echidnas are found only in Australia, Tasmania and New Guinea.

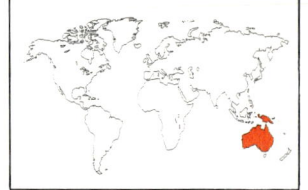

ABOVE The platypus is well adapted to an amphibious life. The ends of the claws on its forefeet are covered by webbing, which the platypus can fold back when on dry ground, leaving the claws free for digging or walking. Its eyes and ears are set in a groove that it can close when underwater.

Courtship in water

Platypuses mate during the months of August to October—springtime in Australia. Courtship starts with a pursuit stage that may last for over an hour. The male follows the female as she swims about, watching her every move. Eventually he tires of this and grasps the female's tail in his mouth. They begin to swim in a circle until the female takes the male's tail in her mouth to close the ring. The male, loosening his grip on her tail, then takes hold of her firmly by the scruff of the neck, and they mate. Some observers say that the male strikes the upper part of the female's body with the spurs on his ankles, presumably without inflicting any damage.

The entire courtship and mating ritual takes place in water and provides a dramatic and noisy spectacle. The female will not always yield to the male, and often has to struggle to keep her head above water so that she can breathe. After mating, which only lasts five to six minutes, the pair may swim around together for about an hour before parting.

Shortly after mating the female starts to build a nursery burrow, lining the nesting chamber with grass or leaves. These are kept damp in order to stop the soft, leathery-shelled eggs from drying out. Once the chamber is prepared she moves in, barring the entrance with a thick layer of earth to keep out other animals. After a gestation of two to three weeks, she lays two—occasionally three—eggs, which she may curl around to incubate. During this time she does not eat. If she does emerge from the chamber and venture out, she will reseal the entrance with a wall of earth when she returns.

From egg to independence

After about 10 days of incubation, the eggs hatch. Like echidnas, young platypuses break open their shells with an "egg tooth" that grows on their jaw bones. They are blind and hairless at birth, and measure about 0.7 in. long. The young open their eyes after 11 weeks; by 4 or 5 months old, their bodies are already covered in soft, brown fur and they are totally weaned. They can then leave the nest and learn to find their own food.

Very young platypuses have about 30 teeth. As the young grow older, their teeth get smaller and eventually disappear, to be replaced by horny plates inside the cheek pouches. The poison gland and ankle spur, present in both male and female young, gradually disappear in the female as she grows older, until only a scar remains.

MONOTREMES CLASSIFICATION

There are only three species in the order Monotremata, the monotremes. They are classified in two families, the Tachyglossidae and the Ornithorhynchidae.

The Tachyglossidae consists of the two species of echidna, also known as the spiny anteaters. The short-nosed (or short-beaked) echidna, *Tachyglossus aculeatus*, is widespread throughout Australia, Tasmania and New Guinea, and occurs in a broad range of habitats. The long-nosed (or long-beaked) echidna, *Zaglossus bruijni,* is much more limited in its distribution; it is found only in the highlands of New Guinea and possibly on the Indonesian island of Sulawesi.

The Ornithorhynchidae consists of one species only: the platypus, *Ornithorhynchus anatinus,* of coastal eastern Australia and Tasmania. It is also known as the duckbill or duckbilled platypus. A semiaquatic species, it occurs in fresh water, in rivers, streams and lakes, requiring places where the banks are soft enough for burrowing.

MARSUPIALS

MAMMALS WITH POUCHES

The marsupials have evolved widely different forms, yet most have one feature in common—their young leave the womb very early and continue their growth in the mother's pouch

MARSUPIALS

Long-nosed bandicoot

Spotted-tailed quoll

Virginia opossum

Thylacine (probably extinct)

Tasmanian devil

Yellow-footed marsupial mouse

Numbat

Marsupial mole

608

MARSUPIALS

The marsupials of Australasia and the Americas are among the most remarkable of all mammals—not only because they have an unusual system of reproduction, but also because of the way the group as a whole has diversified to exploit different habitats and different sources of food.

Altogether the marsupials number some 266 species grouped into 18 families. They range in size from the tiny Formosan mouse opossum, with a body length of less than three inches, to the red kangaroo, which may measure 9 ft. or more from head to tail. Some species are native to Central and South America and one is found in parts of North America, but the majority live in Australia and New Guinea.

From fossil evidence it appears that the marsupials began to follow a different course of evolution from other mammals (the placental mammals) some 100 million years ago. The earliest remains of marsupial animals have been unearthed in North America, but as the placental mammals developed, the marsupials died out in that continent. In South America they survived—facing only limited competition from the placental mammals—and persist as the opossums of today. However, it is in Australia that the group has truly flourished.

Continental isolation

When Australia became isolated from the other continents some 40 million years ago, the marsupials were already well established there. Since the placental mammals had not reached the landmass by then, the marsupials were able to spread into all the habitats the continent had to offer, without competition from other mammals. Some became plant eaters, some insect eaters, and some ate the flesh of other animals.

Each species adapted to its own ecological role. In other continents such roles were taken on by placental mammals. In the course of evolution, many species, such as the marsupial mole and thylacine, came to resemble their Old World counterparts (in this case the mole and the wolf).

What is a marsupial?

In most respects, marsupials are similar to other mammals. They are furry, warm-blooded creatures that suckle their young. The key difference between marsupials and placental mammals lies in the way the young develop in the womb—and outside it.

ABOVE An eastern gray kangaroo "joey" peers out of its mother's pouch. The pouch is more than just a warm refuge or a lazy way for a youngster to get about; for the first few months of life it acts as a second womb, nourishing and protecting the helpless newborn offspring.
PAGE 607 Balanced by its long, heavy tail, a red kangaroo can cover over 33 feet in one bound.

In the vast majority of mammals—including familiar animals such as cats, dogs, sheep, chimpanzees and humans—the unborn young are nourished in the womb by an organ called the placenta. This is attached to the infant by the umbilical cord. Nutrients and other vital substances carried in the mother's bloodstream filter into the placenta and are carried along the umbilical cord to the infant. Waste products from the young pass back down the cord to be disposed of by the mother's body.

The support provided by the placenta allows the young animal to grow within the womb for several months. If necessary, it can reach an advanced stage of development before it is born. A young gazelle, for example, is often capable of running with the herd within hours of birth. Most important, the long period in the womb allows time for the brain—the most complex of all the body's organs—to become

MARSUPIALS

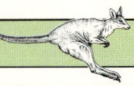

highly developed. A newborn baby may appear ill equipped to face the world, but its sophisticated brain is in full working order.

Compared with young placental mammals, infant marsupials spend very little time in the womb—36 days in the case of the eastern gray kangaroo. Moreover, they do not have the benefit of a placenta. There is a certain amount of nutrient transfer from the wall of the womb through the membranes surrounding the developing young, but this does not compare with the prolonged and comprehensive nourishment provided through the umbilical cord. As a result, the young marsupial is comparatively underdeveloped at birth. By human standards, a newborn kangaroo is little more than an embryo. It is a tiny creature that bears no resemblance to its parent and weighs less than 0.04 oz. when it first encounters fresh air.

LEFT Long claws dug into its mother's fur, a young koala hangs on tight as its parent makes a rare sprint across open country. Koalas spend nearly all their time in eucalyptus trees, feeding on the tough foliage, but it has such poor nutritive value that the animals have to save energy by sleeping for up to 18 hours a day. BELOW The map shows the world distribution of various marsupial species.

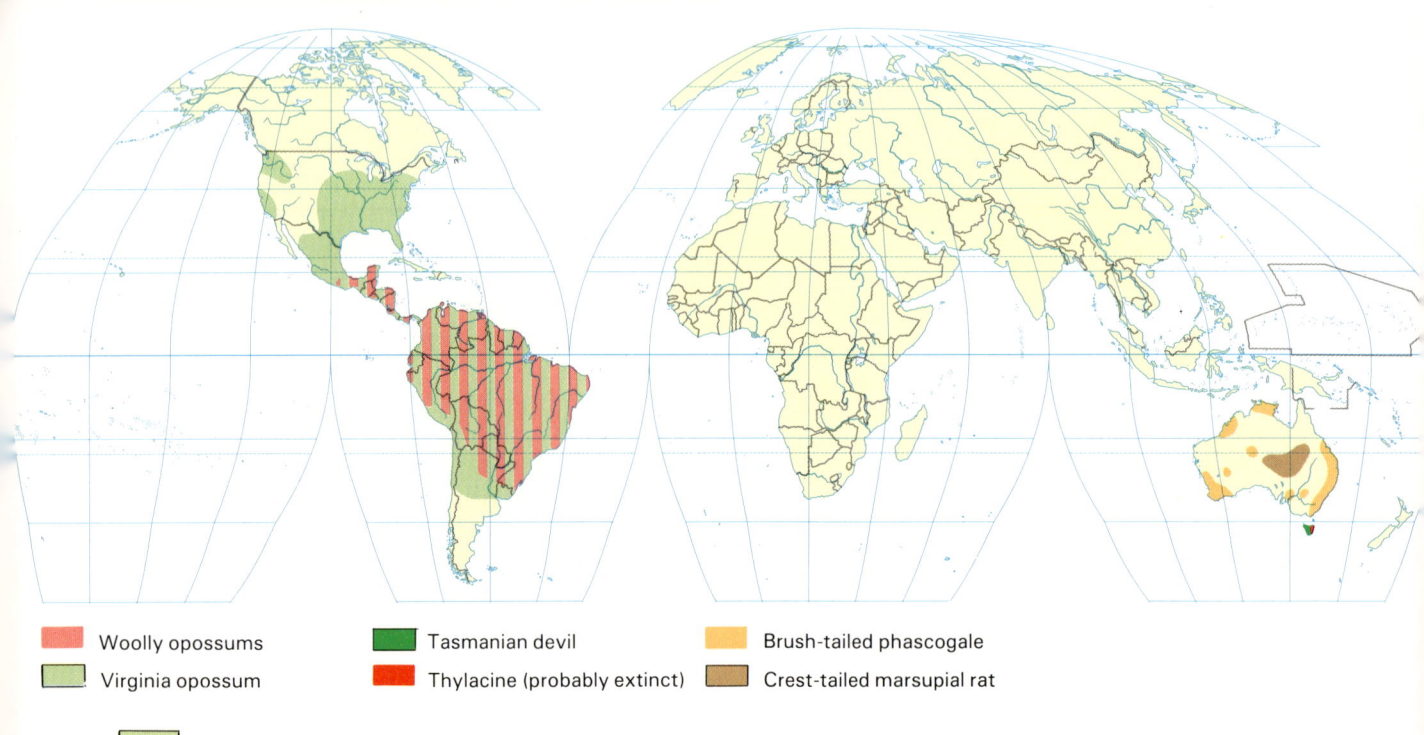

- Woolly opossums
- Virginia opossum
- Tasmanian devil
- Thylacine (probably extinct)
- Brush-tailed phascogale
- Crest-tailed marsupial rat

MARSUPIALS

A young animal in this condition could not survive without its mother's milk, and the tiny kangaroo immediately climbs up its mother's body to reach her teat. In marsupials, the milk supplies all the nutrients and other vital materials that other mammals receive while still in the womb. Since the young kangaroo depends on the teat to the same extent as an unborn lamb depends on the placenta, it becomes just as permanently attached. Drawn into its mouth, the teat enlarges to fill the mouth cavity. In this way the infant becomes connected to its mother almost as effectively as the unborn lamb. The journey from the birth canal to the teat is no more than an interlude between two periods of total dependence.

Protective pouch

As it develops, a newborn marsupial is extremely vulnerable. In many species, such as kangaroos, the mother protects her offspring with a fold or pocket of skin that forms a pouch enclosing the teat and its attached infant. In some ways the pouch could be regarded as a substitute womb; whereas a young placental mammal develops in its mother's womb, a young marsupial does so inside the pouch. However, not all marsupials develop true pouches. Young mouse opossums, for example, are protected only by

TOP A female Virginia opossum may give birth to over 50 young in a litter, but since the mother has only 13 teats, the newborn opossums are faced with a desperate scramble as soon as they are born: any that fail to find a teat when they reach the pouch will die almost immediately.
ABOVE The long, sensitive whiskers on the face of this strikingly marked opossum allow it to feel its way through the branches and undergrowth at night while foraging for food.

MARSUPIALS

LEFT A litter of young opossums clamber over one another to get at their mother's teats. In many opossums, the females have no pouch, and when a mother goes foraging she either carries her young on her back or "parks" them in a nest. When they are very young and totally dependent on their mother, the young opossums have to hang from her teats by their mouths, as she climbs through the branches.

two ridges of skin—as the mother moves around, their tiny bodies dangle from the teats.

Considering its state of development, it is astonishing that a newborn marsupial, such as a kangaroo, ever manages to make the journey to the pouch. As soon as it emerges, the blind, almost helpless infant has to clamber up its mother's belly, clinging to her fur and hauling itself along with its forelimbs. Its mother does little to help, apart from licking a wet trail along her belly fur for the tiny creature to follow. In spite of its difficulties, the young animal generally reaches the pouch within a few minutes of birth. It crawls inside, attaches itself to a teat and begins to take in its mother's milk. It is too helpless to suck the milk actively from the teat; instead, the mother contracts the muscles around the mammary gland and forces milk out through the teat into the infant's mouth.

A marsupial's system of reproduction is an advance over that of egg-laying animals (such as reptiles, birds and monotremes), since it provides the offspring with a long period of development, while protecting it in the pouch. The first stage of development concentrates on physical preparation for the arduous journey from the womb to the pouch. The offspring must be able to move, breathe, feed and digest food before it undertakes the trip. Other mammals, nourished and provided with oxygen automatically by the placenta, focus their earliest development on preparation of the brain. Consequently, the brains of marsupials never achieve the high level of development characteristic in placental mammals.

The opossums

The 75 species of American opossums are extremely varied in size and shape. They are not highly specialized in their way of life or diet. Most are good climbers with mobile, long-toed feet and prehensile tails. They have up to 50 sharp teeth and long, pointed snouts. The majority are natives of the tropical areas of

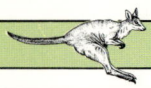

MARSUPIAL CLASSIFICATION: 1

The order Marsupialia consists of 18 families grouped into two suborders: the Polyprotodonta and the Diprotodonta.

The suborder Polyprotodonta contains nine families (although one is probably extinct). They are the American opossums, the Didelphidae; the marsupial carnivores, the Dasyuridae; the numbat, the Myrmecobiidae; the marsupial mole, the Notoryctidae; the bandicoots, the Peramelidae; the bilbies, the Thylacomyidae; the shrew opossums, the Caenolestidae; the monito del monte, the Microbiotheriidae; and, if it still exists, the thylacine, the Thylacinidae.

The suborder Diprotodonta also has nine families. These are the cuscuses and brushtail possums, the Phalangeridae; the ringtail possums, the Pseudocheiridae; the gliders, the Petauridae; the pygmy possums, the Burramyidae; the honey possum, the Tarsipedidae; the wombats, the Vombatidae; the koala, the Pascolarctidae; the rat kangaroos, the Potoroidae; and the kangaroos and wallabies, the Macropodidae.

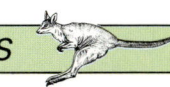

Central and South America, but several are found farther south and the Virginia opossum is expanding its range northward through North America.

Opossums have few enemies, since most predators are discouraged by the unpleasant smell given off by their skin—an odor rather like rancid fat. They can also secrete an evil-smelling substance from two anal glands. If attacked, opossums threaten their enemies by opening their mouths wide and baring their sharp teeth. Opossums have a varied diet, their choice of food depending on the season and the locality. Most species eat nuts and fruit as well as insects and other invertebrates. The larger species also eat small mammals and reptiles. Opossums frequently eat carrion (the flesh of dead animals) and will even rummage through garbage dumps.

The gestation period for opossums is very short, usually lasting two weeks or less. The number of young that are born may greatly exceed the number of available teats in the mother's pouch—those that cannot find a vacant teat soon die. The front toes of the tiny young opossums are equipped with claws to enable them to crawl up their mother's belly into the pouch, but these disappear as the young animals grow older. From 3 to 18 offspring can develop in the pouch. At first they stay attached to their mother's teats, but later they only come to the teats to feed. They spend the rest of their time either clinging to her back or lying hidden in a nest while she goes out to forage.

ABOVE LEFT Clinging to the bark with its long toes, a Virginia opossum peers out from a high, forked branch. Many opossums are excellent climbers, and take to the trees both to find food and to avoid predators.

ABOVE Disturbed during its nighttime search for food, an angry mouse opossum opens its mouth threateningly and bears its teeth. When mouse opossums are roused, their large, delicate ears crumple into tight folds.

The Virginia opossum

The best known of the American opossums is the Virginia opossum. It ranges from southeast Canada, through the eastern half of the USA, to Central America; it has also been introduced to the west coast of the USA. It is about 31 in. long, though almost half of this length is tail, and it has a gray-brown or reddish coat and a white face. Its luxuriant fur consists of a thick undercoat concealed beneath longer, protective hairs with white tips. The end of its nose is naked pink skin, while its ears and the areas around its eyes are dark. It has a long, scaly tail covered in hair at the base.

The Virginia opossum is generally solitary—except during the breeding season—and is most active at

MARSUPIALS

American opossums range from Patagonia through Central America to the United States and southern Canada.

night. It spends its day concealed in a nest up a tree, in a hollow tree trunk or in a rock cavity. Its nest is lined with leaves, grass or any other material that the animal can collect and carry in its mouth. It is an excellent climber, with a muscular, prehensile tail. The tail acts as a fifth limb, providing extra grip and stability as the opossum clambers among the branches in search of birds, small mammals, eggs and fruit. In more northerly areas, some of the opossums become inactive during the cold winter months.

Virginia opossums are well known for their habit of "playing possum," or feigning death, when danger threatens. Their passive defense strategy is not uncommon among small animals, but it is particularly well developed in the Virginia opossum. The animal curls up on the ground and stays absolutely motionless. The behavior is a nervous reaction, comparable to fainting, and is brought about by harmless, semiparalyzing substances that are released into the bloodstream as a response to high levels of stress. These substances cause the opossum's muscles to contract and maintain the apparent "rigor mortis" until the danger has passed and the stress level drops back to normal. Though the reaction may last only half a minute, in some cases the animals remain motionless for as long as six hours.

Fierce fighters

The mouse (or murine) opossums of South America are all small, mouse-like marsupials. Some, such as the ashy mouse opossum, are mainly tree dwellers. They are good climbers with long, slender prehensile tails. Others spend more time on the ground, are less agile and have shorter tails that can become swollen with fat as an energy store.

The water opossum, or yapok, ranges from southern Mexico to northeast Argentina. It has become highly adapted to life in the water. Measuring 11-16 in. in body length, it has a gray and black back and a white underside. Its tail is 12-17 in. long and is furred at the base. Uniquely among opossums, it has webbed hind feet that, together with its streamlined body and fine, dense, water-repellent fur, make it a very efficient swimmer. The female's pouch can be sealed by strong constrictor muscles to form a watertight chamber for the young opossums when their mother goes diving for food. Males also possess pouches into which they withdraw their genitals when they are swimming.

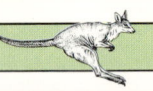

TOP AND ABOVE The Virginia opossum has increased its range considerably during the last 200 years. Before 1800 it was limited to Panama, southern Mexico and the southeastern United States. Today it is found as far north as Canada and has colonized the American west coast following its introduction to California in 1890. Its success is largely due to its adaptability and willingness to live near man. It has a varied diet and has often been found searching for food scraps around people's homes.

MARSUPIALS

ABOVE **Diving with the grace and speed of an otter, a water opossum curves through the water to catch a fish. With its webbed hind feet and watertight pouch, it is the only marsupial that is fully adapted to a semi-aquatic life.**

ABOVE RIGHT **The ashy mouse opossum is the largest member of the mouse opossum genus and also one of the most agile. It spends most of its time foraging in the trees and will often swing by its strong prehensile tail to reach its food.**

Australian carnivores

The marsupial carnivores are a large and flourishing group of animals living in Australia, Tasmania and New Guinea. They include true carnivores that prey on other vertebrates, such as rodents and reptiles; and insectivores that feed on insects, spiders, worms and similar invertebrates.

Marsupial carnivores include tree dwellers, ground dwellers and species adapted for running or jumping. Many bear a strong resemblance to unrelated placental mammals, having evolved in similar ways to cope with the same ecological conditions. At the time when the marsupial carnivores were starting to diversify, there were no placental mammals in Australia, and the marsupials filled the places in the food chain that were left vacant. As a result, the evolutionary process has produced a range of

MARSUPIAL CLASSIFICATION: 2

Didelphidae

The 75 species of American opossums belong to the family Didelphidae. They range over South and Central America and parts of North America, and occur in most types of habitat. Most species belong to the subfamily Didelphinae, which is divided into eight genera. They include the Virginia or common opossum, *Didelphis virginiana*, of North and Central America; the water opossum or yapok, *Chironectes minimus*; the little water opossum, lutrine or thick-tailed opossum, *Lutreolina crassicaudata*; the 14 species of short-tailed opossums (genus *Monodelphis*); and the 47 species of mouse opossums (genus *Marmosa*).

The other subfamily is the Caluromyinae, members of which range from southern Mexico to northern South America. They are the bushy-tailed opossum, *Glironia venusta*; the black-shouldered opossum, *Caluromysiops irrupta*; and the three species of woolly opossums belonging to the genus *Caluromys*.

MARSUPIALS

Spotted cuscus

Common wombat

Bennett's tree kangaroo

Ring-tailed rock wallaby

Sugar glider

Eastern gray kangaroo

Koala

616

MARSUPIALS

TOP RIGHT **The black-tailed phascogale, a shrew-like marsupial carnivore of Australia, feeds on insects rather than animal flesh.**
CENTER RIGHT **In the stony desert of central Australia, a kowari devours a locust. The kowaris and phascogales are among the** smallest of the marsupial carnivores, weighing from 2 to 11 oz.
BOTTOM RIGHT **A kowari is shown: standing on tiptoe supported by its tail as it curiously sniffs the air (A); in a vigilant posture (B); and curled up in deep sleep (C).**

marsupials that, in appearance and behavior, resemble Old World mammals such as shrews, rats, cats, dogs and even bears.

Mostly active by night, the marsupial carnivores tend to be aggressive. Observations of captive animals have indicated that the marsupial carnivores have voracious appetites; many are able to eat more than their own weight in food in one day—although whether they do this in the wild, where food is not so easy to come by, is open to question. Like opossums, they have many teeth, including well-developed, sharp canines. The family includes the planigales, the phascogales, the quolls and the Tasmanian devil.

The diminutive planigales

The planigales include some of the smallest of all the marsupial species; the bodies of some adults are no more than two inches in length. They have flattened skulls that, together with their small size, allow them to pass through tiny crevices—in the same way as many small rodents, bats and reptiles. These mammals are insectivorous, but being so small they are frequently outsized by their own prey.

There are two species of phascogale, also known as wambengers or tuans. The most widespread is the brush-tailed phascogale, which occurs in many parts of Australia. It is 6-9 in. in length with a black tail stretching a further 7-9 in. Both the brush-tailed phascogale and the smaller red-tailed phascogale have thick, bluish gray coats and white bellies. In both species the end of the tail is bushy and the long, bristly hairs rustle loudly when they are rubbed. Phascogales have a rudimentary pouch that develops only during the mating season.

A miniature bear

The Tasmanian devil lives only on the island of Tasmania, south of mainland Australia. It occurs throughout the island, except in places where the

617

MARSUPIALS

ABOVE AND ABOVE RIGHT Looking like a cross between an otter and a puppy, the Tasmanian devil has gained a reputation with many settlers for being a fierce glutton. After it has fed on a carcass, little is left except the largest bones. Although it is a predator and will kill to eat, it relies mainly on carrion for most of its food.

vegetation cover has been extensively cleared. The Tasmanian devil was once very rare, owing to disease, habitat destruction and extermination by settlers. In the early 1900s it came near to extinction, but it has since recovered and is even common in some areas.

The Tasmanian devil is a thick-set animal, just under three feet long, and resembling a small bear. It has a short, broad head, short limbs and a stumpy tail. Mostly black, it has a distinctive white stripe across its throat, and some individuals have white spots on their head and flanks.

The ground-dwelling, carnivorous Tasmanian devil is well known for being ferocious, and it has an impressive defense display of loud snarls and gnashing teeth. However, it is not an efficient predator and seems to live mainly on carrion, preferring large carcasses such as sheep, wombats and wallabies. Increases in the amount of carrion from livestock and animal trapping have probably helped it recolonize areas and boost its population. If it is available, the Tasmanian devil will take live prey, catching small wallabies, kangaroo rats, birds, reptiles, amphibians and even fish.

The end of the line

The thylacine, or Tasmanian wolf (sometimes referred to as the Tasmanian tiger), is probably extinct, although unconfirmed sightings are still reported from time to time in its last stronghold—the most remote parts of Tasmania. Most of the sightings are probably cases of mistaken identity and, even if there are a few thylacines still alive, the population must be so small that it cannot survive for long.

At first glance a thylacine looked (or looks) like a dog, with a dog-like head, legs and build. Indeed, its scientific name is equivalent to "pouched dog with a wolf's head." The thylacine differed from dogs in having an extended rump, which projected behind the thighs, and a series of black tiger-like stripes across the hind part of its back. Also, it was a marsupial and therefore quite unrelated to any species of dog. However, it had come to resemble a dog because it lived in the same habitat and had the same way of life. Like the dog, it was a predator that captured its prey both by stealth, tracking down and surprising its victim, and by running down animals in the open. However, as far as is known it did not run in packs, but preferred to hunt alone or in pairs.

Fossil remains indicate that three thousand years ago the thylacine occurred all over mainland Australia. However, it had disappeared from there many centuries before the first European settlers arrived, probably because of competition with dingoes. The thylacine flourished in Tasmania until it came into contact with settlers—particularly sheep farmers, who regarded the animal as a pest. The government actually offered a reward for each thylacine killed, and between 1888 and 1909 a total of 2268 were exterminated as a direct result of this policy. Such slaughter, coupled with a disease epidemic, sealed the fate of the thylacine, and the last authenticated specimen died in Hobart Zoo in 1934. The last definite record of one killed in the wild dates from 1930.

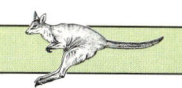

MARSUPIAL CLASSIFICATION: 3

Dasyuridae

Many of the carnivorous marsupials of Australasia belong to one family—the Dasyuridae, known as the marsupial carnivores. The family is divided into 18 genera, containing at least 51 species (the precise classification of some of these animals is still under debate). They range throughout mainland Australia, Tasmania and New Guinea. They include the five species of planigales (genus *Planigale*); the ten species of antechinus (genus *Antichinus*); the kultarr, *Antechinomys laniger*; the two species of phascogales (genus *Phascogale*); the kowari, *Dasyuroides byrnei*; the quolls or native cats (genus *Dasyurus*); the dunnarts and their allies (genus *Sminthopsis*); the two species of ninguai (genus *Ningaui*); six species of marsupial mice (genera *Murexia*, *Myoictis*, *Neophascogale* and *Phascolosorex*); and the Tasmanian devil, *Sarcophilus harrisii*, now found only in the remote interior of Tasmania.

Thylacinidae

The thylacine or Tasmanian wolf, *Thylacinus cynocephalus*, belongs to a family of its own, the Thylacinidae. Unfortunately, the thylacine has not been seen in the wild in its native Tasmania since the 1930s, and is probably extinct.

Notoryctidae

The marsupial mole, *Notoryctes typhlops*, also belongs to a separate family, the Notoryctidae. It spends most of its time underground, occasionally coming up to the surface, and is found in the arid, sandy soils of central Australia.

Myrmecobiidae

Another distinctive marsupial that has proved difficult to group with others is the numbat, *Myrmecobius fasciatus*. Accordingly, it too has been classified in a separate family, the Myrmecobiidae. Once common throughout southern and central Australia, it is now found only in the forests of the southwest.

TOP Stuffed museum exhibits are probably all that remain of the thylacine or Tasmanian wolf, the largest of the marsupial carnivores. Although unproven sightings are still reported from the most remote areas of Tasmania, all the evidence suggests that the last of these dog-like predators died out in the late 1930s.
ABOVE Ears cocked, a numbat stands on its hind legs to survey its surroundings before continuing the search for termites.

MARSUPIALS

ABOVE The numbat's elongated snout and strong claws make it well equipped to catch the termites that make up most of its diet. Once it finds their nests under stones and dead wood, it digs rapidly down through the galleries, licking up the termites with rapid flicks of its long, sticky tongue. The numbat was once widespread across Australia, but is now restricted to the forests of southwest Western Australia.

Marsupial moles

The marsupial mole is a unique species: it bears a strong resemblance to the common European mole and the African golden mole, but in fact is unrelated to them. It is much the same size (about 6 in. long head to tail) and has developed similar adaptations for an underground, burrowing way of life. It also has a similar diet, consisting largely of insects—especially beetle and moth grubs. It differs from them in that the female has a marsupial pouch for carrying her young.

Like the European mole, the marsupial mole has a compact, cylindrical body and short limbs adapted for moving earth. Its front feet are each equipped with two long, broad claws that act as shovels, and all four feet are flattened. It has tiny eyes and ears hidden by its silky, whitish to golden-colored fur, and a horny shield on its snout. The mole burrows by thrusting this reinforced snout through the soil while it excavates with its large front claws. As the mole shifts forward, it uses its hind feet to lift the freshly dug soil and throw it backward. The vertebrae of the mole's neck are fused to form a rigid structure that may have adapted to lend extra support to the head, increasing the thrusting power of the snout.

The sideways sweep of the marsupial mole's front claws—similar to the arm movements of a breaststroke swimmer—gives the impression that it is swimming through the soil. The marsupial mole does not dig permanent burrows, unlike the European mole. It restricts its mining activities to small tunnels a few inches below the surface. It often comes up above ground and shuffles along at a frantic pace with its nose pointing downward in search of food. The mole leaves behind a distinctive winding trail, made by its belly, feet, and stumpy tail. After staying on the surface for a while, the mole will plunge its snout into the ground and start to dig a new tunnel.

A striped coat

The numbat is a marsupial species in a family on its own. Measuring some 14 in. long, the numbat has a coat of short, coarse, reddish gray hair with distinct white stripes across the hind part of its back. It also has a dark stripe on each side of its muzzle, running from nose to ear. It has a long, bushy, squirrel-like tail, and its body seems to be curiously out of proportion—the hindquarters are much bulkier than the forequarters. Female numbats do not have a pouch. Once attached

---

MARSUPIAL CLASSIFICATION: 4

Peramelidae

The 15 species of bandicoots within the family Peramelidae are grouped into six genera. They occur throughout Australia, Tasmania, New Guinea and nearby islands. There are three species of Australian long-nosed bandicoot (genus *Perameles*); three species of spiny bandicoot (genus *Echymipera*); four New Guinea long-nosed bandicoots (genus *Peroryctes*); three species of short-nosed bandicoots (genus *Isoodon*); the mouse bandicoot, *Microperoryctes murina*; and the Ceram Island bandicoot, *Rhynchomeles prattorum*.

Thylacomyidae

Since the lesser bilby or lesser rabbit-eared bandicoot, *Macrotis leucura*, is probably now extinct, the family Thylacomyidae may consist of only one species: the greater bilby, greater rabbit-eared bandicoot or dalgyte, *Macrotis lagotis*, of the Australian desert.

MARSUPIALS

RIGHT A tiny, fragile honey possum clings to one of the flowers that provide it with its main food — nectar and pollen. These marsupials are found only on the heathlands of southwest Australia where there are flowers in bloom all year round.

BELOW A young long-nosed bandicoot climbs into the large, rear-opening pouch of its mother.
PAGES 622-623 Two young sugar gliders, hanging upside-down from their mother, cling tightly as she moves slowly along a branch.

to their mother's teats, the offspring are protected only by her fur. When they are old enough to be left on their own for short periods, they are placed in a nest, either in a hole in the ground or in a hollow tree, where they stay until they are fully weaned.

The numbat's name is Aboriginal and means "eater of ants" — not a strictly accurate name. It is really a termite eater and will only eat ants if they are in the way when it is feeding. Predatory ants will rush in as soon as the numbat exposes a termite colony. It forages busily along the forest floor, probing among the fallen wood and leaf litter to locate termite runs near the ground surface, scooping up the insects with its long, slender tongue. It will occasionally climb trees if food is likely to be found under the bark.

The numbat's forefeet are equipped with large claws, which it uses to dig out the termite nests and rip off bark. Its elongated, tapering snout also makes a useful lever for removing large stones and branches. Unusual for a marsupial, the numbat does all its foraging by day, and may even bask in the sun during the summer. It spends the night holed up in a burrow or inside a hollow log, which it lines with grass and leaves.

"Pig rats"

The bandicoots of Australia and New Guinea are rat-like marsupials that root around in the earth for their food, using their claws and long, pointed snouts to dig out insects and worms. The combination of their appearance and behavior has earned them their common name, bandicoot, which means "pig rat."

Bandicoots run with sudden and frequent changes of direction. Like kangaroos, their hind legs are more developed than their forelegs, and each hind foot ends in five toes. The three inner ones are long, while the two outermost ones are rudimentary. The extinct pig-footed bandicoot was specialized for running, resting its weight on one toe of the forefoot and two toes of the hind foot, but the surviving species are not so highly adapted.

The body length of bandicoots, excluding the tail, varies from 6 in. in the mouse bandicoot to the 22 in. or more of the giant bandicoot. All species are nocturnal, spending the day under cover and foraging at night for their food. Their diet includes a variety of insects, spiders and other small invertebrates, as well as fungi and roots.

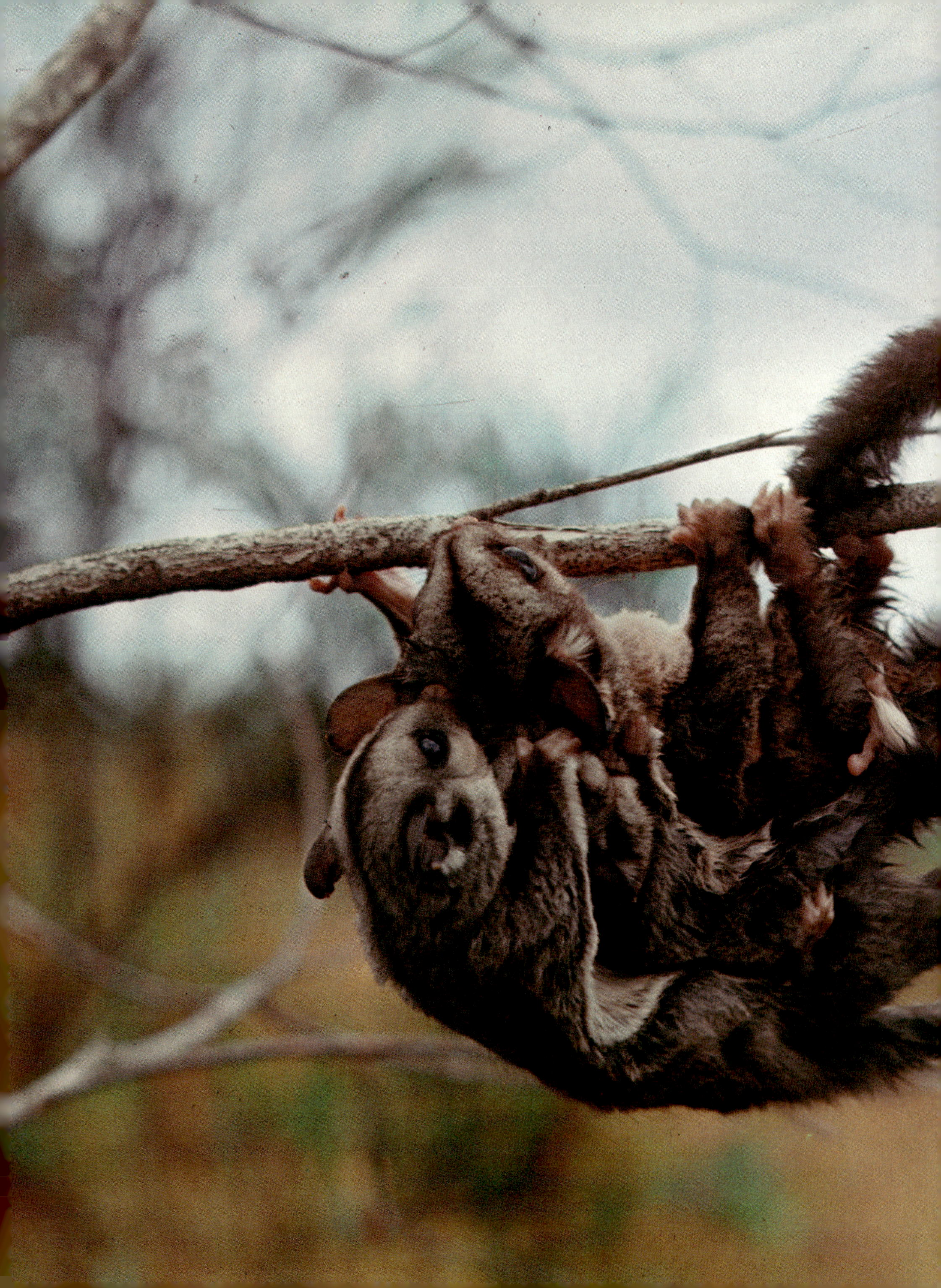

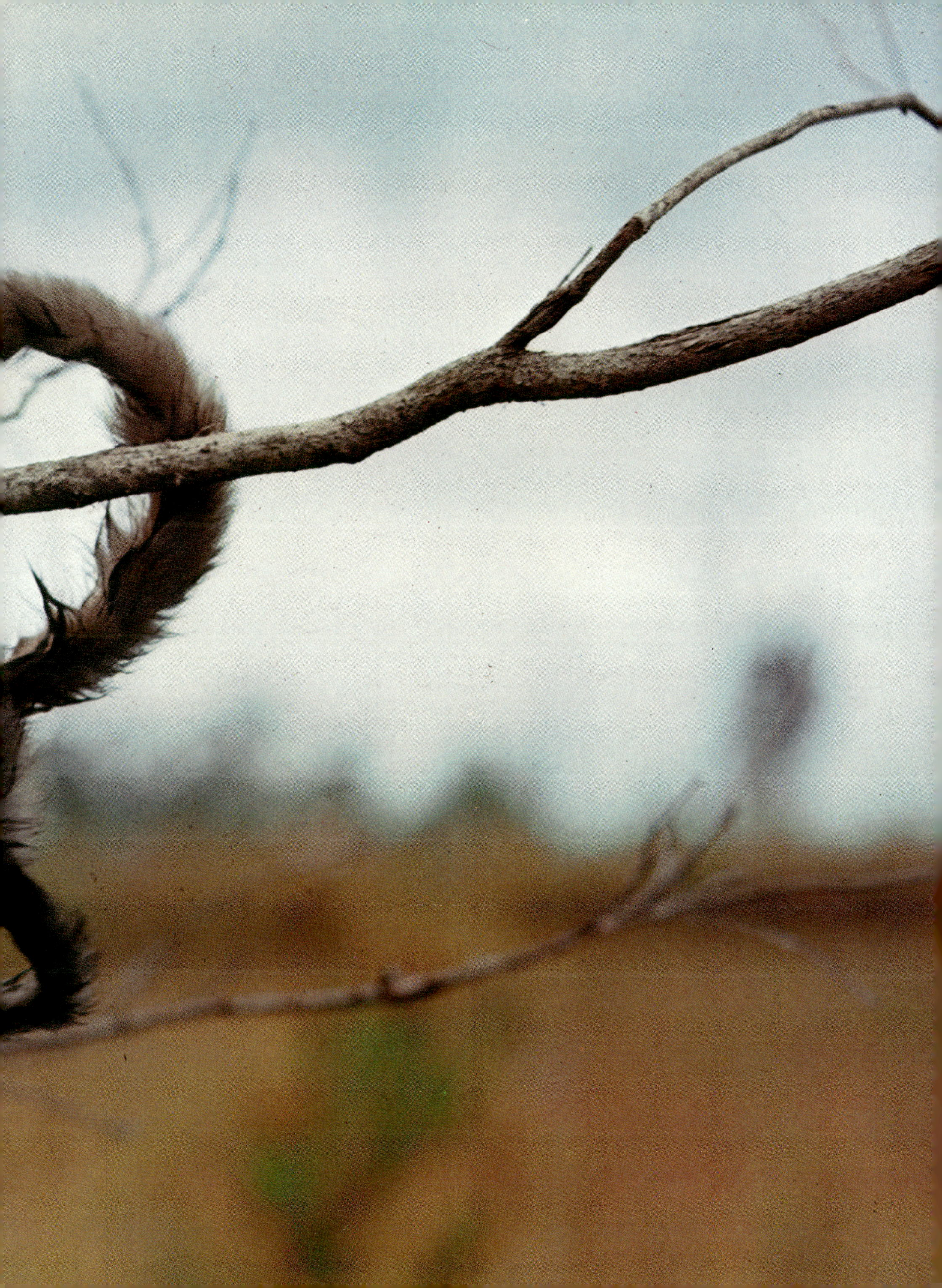

Fast reproduction

The bandicoot reproduces at an extremely fast rate. Bandicoots have very short gestation periods (about 12 days), and the young stay in the pouch for about 50 days. After a further 10 days of semi-independence, they are weaned and leave to fend for themselves. The mother is then ready to produce another litter. The quick development of the young is aided by a simple form of placental feeding within the mother's uterus that gives richer nourishment. The young are consequently more developed when they are born than the young of other marsupials. They mature early (within 90 days), and the young females are ready to begin breeding at three months old.

A host of enemies

The bandicoot's main enemies are man, and the dogs, cats and foxes that man has introduced. Bandicoots have always been hunted by the Aborigines for their meat and hides, but they are now persecuted for causing damage to crops and flower beds in gardens and orchards, in their search for grubs and worms. Such persecution is one reason why bandicoots are declining faster than most other marsupials. Another reason is the destruction of protective ground cover by grazing sheep and cattle, and by wild rabbits. (Rabbits are another animal introduced into Australia by European settlers.) In areas of low rainfall, where huge stretches of land are given over to the grazing herds, bandicoots are in real need of protection.

The long-nosed bandicoot has a reddish gray coat and a rat-like appearance, small ears and an extended snout. Its jaws have many small, pointed teeth typical of insect-eating animals. Normally found in thickly vegetated regions of Australia and Tasmania, it also occurs near towns, where it takes advantage of the wealth of food available in parks and gardens. During the day, it rests under bushes or in ditches, emerging at night to feed on insects, worms, small mammals, birds and reptiles, and sometimes plants. It will occasionally burrow to find food or to hide from predators, but despite this, dingoes and foxes find it an easy prey.

Shrew opossums

The shrew opossums are well named, for these small marsupials are similar to shrews in appearance. They have long, pointed snouts well equipped with sensitive whiskers and a pair of lower incisors that project forward. Their ears are quite large, though partly hidden by thick fur, and they have small, relatively ineffective eyes and hairy tails.

Fossil evidence indicates that during the Tertiary period, some 20 million years ago, there were 7 genera of shrew opossums, and these were very common all over South America. The 3 surviving genera are now found on the western side of the Andes up to 14,100 ft. above sea level. They hunt by night, killing insects, worms, small mammals and reptiles with their long incisor teeth.

The possums

Three very similar groups of tree-dwelling marsupials exist today: the brushtail possums, the rare scaly-tailed possum and the cuscuses. (The possums of Australia are a quite distinct group from the opossums of the New World.) The cuscuses are natives of New Guinea, although some species occur elsewhere (on the Cape York Peninsula of Queensland, Australia,

---

MARSUPIAL CLASSIFICATION: 5

Caenolestidae

The shrew opossums of South America make up a small family—the Caenolestidae—with seven species grouped into three genera. The main genus is *Caenolestes* with five species: the gray-bellied shrew opossum, *C. caniventer*; the blackish shrew opossum, *C. convelatus*; the Colombian shrew opossum, *C. obscurus*; the Ecuadorean shrew opossum, *C. fuliginosus*; and Tate's shrew opossum, *C. tatei*. The other two genera contain one species each: the Peruvian shrew opossum, *Lestoros inca*; and the Chilean shrew opossum, *Rhyncholestes raphanurus*.

Microbiotheriidae

The sole member of the family Microbiotheriidae is the monito del monte, *Dromiciops australis*, which lives in the cool, humid forests of southcentral Chile in South America. It is a small marsupial, with a long, thick tail.

MARSUPIALS

ABOVE A spotted cuscus threads its way through the dense foliage. This sloth-like marsupial is very variable in color and coat pattern; it may be uniform white or gray, or marbled with dark gray or reddish brown spots or blotches. The patterned forms are often males, whereas females tend to be gray or white all over. ABOVE RIGHT Holding onto a branch with its long-clawed feet, a white form of the spotted cuscus appears like a ghost in the artificial light of the camera.

and on some Indonesian islands); the possums are found in Australia.

These three groups—called "phalangers" because of their long toes or "phalanges"—are specialized for life in the trees. Each foot has five toes bearing strong claws. The big toe on the hind foot, and both the first and second toes on the forefoot, are opposed to the others (the fronts of the toes face each other and can be pressed together). Their tails are long and prehensile—able to wrap around a branch. All these features ensure that they keep a good grip as they climb among the trees.

The phalangers are most active during the evening and at night, foraging mainly for vegetarian food such as fruit and leaves, but supplementing this with animal food such as insects and birds' eggs. Some species might take small reptiles, birds and mammals.

MARSUPIAL CLASSIFICATION: 6

Phalangeridae

The family Phalangeridae, the brushtail possums and cuscuses, consists of 14 species, grouped into three genera. There are three brushtail possums of the genus *Trichosurus*: the common brushtail possum, *T. vulpecula*, found all over Australia; and the mountain and northern species, *T. caninus* and *T. arnhemensis*.

The genus *Phalanger* comprises 10 species of cuscuses or phalangers. Eight of these are native to New Guinea, including the ground cuscus, *P. gymnotis*; the silky cuscus, *P. vestitus*; and the mountain cuscus, *P. carmelitae*. Small populations of two species also occur in northern Queensland: these are the spotted cuscus, *P. maculatus*, and the gray cuscus, *P. orientalis*. Two other species occur as far west as Sulawesi: these are the Sulawesi cuscuses, *P. ursinus* and *P. celebensis*.

The scaly-tailed possum, *Wyulda squamicaudata*, occupies a genus of its own and is found in the Kimberley region of Western Australia.

MARSUPIALS

Cuscuses and brushtail possums occur over most of Australia and on many of the islands to the north.

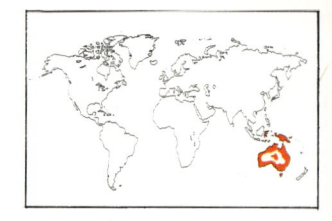

A familiar marsupial

The common brushtail possum, probably the mammal most often encountered by Australians, is widespread throughout Australia and Tasmania. Once restricted to the forests, it can now be found in open, rocky areas and even in towns, where it finds shelter beneath the roofs of old houses and outbuildings. It is unusual among Australian marsupials in that it has managed to extend its area of distribution despite the presence of man, and despite the fact that its soft pelt is highly valued by the fur trade.

The common brushtail is not unlike a large squirrel, with a body almost 24 in. long, soft, dense fur and a long, thick, prehensile (grasping) tail. The coat markings vary between the female and the male: the female is grayish, while the male is redder or tawny-yellow. It is a shrewd hunter and fighter, with opportunistic eating habits that probably go some way toward explaining its success. It can also breed throughout the year, producing one offspring at a time; the offspring leaves the pouch at four to five months old and hangs onto its mother's back until it has been weaned.

The most elusive of all these possums is the scaly-tailed possum, which was quite unknown before 1917 when it was discovered in the remote northern Kimberley region of Western Australia. It is named for its naked, prehensile tail, which has a knobbly, scale-like surface. Exclusively vegetarian, it feeds on the leaves and flowers of eucalyptus trees.

The cuscus

Cuscuses are larger and heavier than the brushtail possums. There are 10 species, of which the most common is the spotted cuscus, found in the rain forests of New Guinea, northern Queensland, and the adjacent islands. It is a bulky animal, about three feet long with a very thick coat and a prehensile (grasping) tail. It has large, slightly protruding eyes, small ears and long, curving claws. The different coat patterning of the sexes is most marked in this species. The male is generally yellow or white with dark spots, while the female is usually pale all over; both sexes tend to have golden-yellow faces.

Marsupial gliders

Among the marsupials, a number of species are able to glide through the air from tree to tree. To do so they have a membrane of skin (known as a patagium)

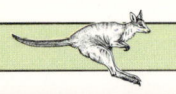

ABOVE A common brushtail possum demonstrates its agility as it makes its way down the trunk of a pine tree. One of the most adaptable of the Australian marsupials, the common brushtail is equally at home in the city suburbs and the remote outback. It feeds on a wide range of foliage including eucalyptus, garden plants and even grass. In large numbers, they can strip whole trees of their leaves. Deprived of the life-giving energy from the leaves, the trees are then left to die.

MARSUPIALS

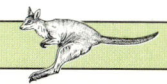

RIGHT At rest, the sugar glider resembles an edible dormouse, only the wrinkled flap of skin along each flank giving any hint of its spectacular gliding abilities. Sugar gliders can glide for distances of over 164 ft. They live exclusively in trees and are sociable mammals, often building communal nests that may last for several seasons.

MARSUPIAL CLASSIFICATION: 7

The pygmy possums, gliders and ringtail possums are all forest-dwelling marsupials, and include the nine species of "flying" marsupial.

Burramyidae

There are seven species of pygmy possum in the family Burramyidae, grouped into four genera. They include the feathertail glider or flying mouse, *Acrobates pygmaeus*, of eastern Australia; the feathertail possum, *Distoechurus pennatus*, of New Guinea; the four species of pygmy possum in the genus *Cercartetus*, which occur in Tasmania, New Guinea and much of coastal Australia; and the mountain pygmy possum, *Burramys parvus*, of southeast Australia.

Petauridae

The gliders of the family Petauridae consist of seven species in three genera. There are two striped possums of the genus *Dactylopsila*, which live in New Guinea and northern Queensland. Four species belong to the genus *Petaurus*, including the sugar glider, *P. breviceps,* which is widespread from Tasmania to northwestern Australia and New Guinea. The remaining species is the very rare Leadbeater's possum, *Gymnobelideus leadbeateri*, that occurs in Victoria.

Pseudocheiridae

Ringtail possums belong to the family Pseudocheiridae. There are 16 species; 15 of these belong to the genus *Pseudocheirus*, which includes the common ringtail possum, *P. peregrinus*, widely distributed throughout the wooded regions of Australasia from Tasmania to New Guinea. The other genus contains only one species, the greater glider, *Petauroides volans*, found in eastern Australia.

connecting the forelimbs with the hind limbs to form a primitive wing. The ability to glide has been developed independently by members of three marsupial families—the pygmy possums, the gliders and the ringtail possums—that inhabit Tasmania, coastal Australia and New Guinea.

The feathertail glider is one of the pygmy possums. At only 3.5 inches in length, it is one of the smallest of the marsupials and certainly the smallest of the gliders. Also known as the flying mouse, it lives among the treetops in the eucalyptus forests of northern Australia, and rarely comes down to ground level. Instead, it glides among the trees by leaping into the air with its legs outstretched, spreading the patagium that links the wrist and the knee on each side. The edge of the membrane is fringed with long hairs to increase its area, and the animal's tail also has a fringe of hair that turns it into a movable, feather-like extra wing. Such a tail both increases the wing area and provides a means of steering during the glide.

Feathertail gliders are gregarious animals, often living in large family groups. They forage by night, feeding on nectar and pollen from flowers, and on the occasional insect. They often live very close to humans without being noticed, since their whole lives are spent high in the treetops. Indeed, when a tree is

THE KOALA
— A SPECIALIZED FEEDER —

Dumpy and tailless, with a large head, a black, shiny button nose and two round, fluffy ears, the koala looks like the perfect pet for a child. However, few animals are less suited to living in captivity. Even zoological parks, with all the resources at their disposal, find it difficult to provide the koala with all its needs.

Measuring about 28-31 in. long, the koala may reach a weight of 33 lbs. on a diet that consists almost exclusively of the leaves and buds of a few species of eucalyptus. Koalas are extremely fussy about which leaves they will eat, and in which season. During winter, for example, when the foliage of the manna gum tree—a favorite of the koala—produces a high degree of poisonous prussic acid, the koalas search for eucalyptuses with a lower toxic level. The eucalyptus leaves contain large quantities of essential oils—including the familiar eucalyptus oil used in cough drops—and these give the animal a strong scent and help protect it from parasites. The water contained in the leaves is also an important source of moisture for the koala, and in all but the hottest weather it obtains all the liquid it needs from the foliage.

A tree-bound existence

A koala rarely comes down to the ground unless it needs to cross the ground to reach an isolated eucalyptus tree. It stays in the trees day and night, eating, sleeping, mating and raising its young among the branches. It is most active at night, beginning its browsing at dusk and retiring to a comfortable fork in the tree to sleep. It is well equipped for this tree-dwelling way of life. On each forefoot, the first two digits are opposable to the other three, giving the koala, in effect, two thumbs. On the hind feet, the first digits are opposable to the other four.

These features, in addition to the long, pincer-like claws on all fingers and toes, provide the animal with an excellent grip on the smooth bark. The koala climbs vertical tree trunks by grasping the bark with its forelimbs, then lifting both hind limbs together, digging its claws in and pushing itself up.

628

Koalas live alone in neighboring territories, but during the summer breeding season (October to February in the Southern Hemisphere), an adult male will mate with any receptive females whose territories overlap his own. Meanwhile, he chases off other males who enter his territory, and warns others to keep away by making a repeated harsh roaring call. Other adult males respond with the same roars, and the mating season can be a noisy time. The female usually stays silent, although she will produce a wailing cry of distress if a male attempts to mate with her when she is not ready. Young males, too, will give a similar distress call when threatened by adult males.

ABOVE RIGHT Of the 350 different species of eucalyptus that exist, koalas will eat the leaves of only 20.
RIGHT As long as they stay in trees, koalas are comparatively safe, but on the ground they are vulnerable to foxes and dogs.
ABOVE A mother koala tries to keep its restless youngster under control.
FAR LEFT The koala's diet is so nutritionally poor that the koala has to sleep for as much as 20 hours each day to conserve its energy.
LEFT A young koala is dependent on its mother for about five months after leaving the pouch. When born, the koala is about the size of a grape.

MARSUPIALS

ABOVE **The mouse-like feathertail glider is the smallest of the gliding marsupials, and one of the most acrobatic. As well as the gliding membrane joining its limbs, it has a flattened tail with a fringe of long hair that forms a feathery aerofoil; this allows the glider to steer itself through the air and land exactly where it chooses. It feeds on nectar sipped from flowers with its brush-tipped tongue.**

chopped down in a garden, a whole colony of feathertails may appear as if from nowhere, and hastily move off to find a new hiding place. They build nests in the trees and produce an average of four offspring in each litter.

Airborne family

Most of the flying species belong to the family appropriately known as the gliders. The most common of these animals is the sugar glider, which lives in the open woods and forests of Australia, New Guinea and Tasmania. It is a much larger animal than the feathertail glider, the male measuring some 7 in. long, with another 8 in. of tail.

The sugar glider has a thick, gray coat with a black stripe running down its spine, and a bushy, squirrel-like tail. Its tail is not flattened like that of the feathertail glider, but is nevertheless used as a rudder when the animal leaps through the air. Sugar gliders are communal: each individual is marked with the scent of the dominant male in the group, and his scent is also used to mark out territory. Sugar gliders feed on sap—preferably gum from the wattle tree—together with fruit, nectar and even insects captured during flight. They are aggressive animals and will sometimes attack small mammals that come too close.

The greater glider

The largest of the "flying" marsupials is the greater glider, a member of the ringtail possum family. It is so large—up to 20 in. long excluding its tail—and has such an extensive patagium that it can cover considerable distances in a single glide. But it does not have the maneuverability of smaller species. Compared to the sugar glider it has an odd-shaped patagium, for it extends from the elbow of the forelimb to the ankle of the hind limb; in flight its outline is almost triangular, giving it a "delta-winged" appearance.

The greater glider lives in the rain forests and humid woodlands of eastern Australia, and spends most of its time alone or in pairs. It is a specialist leaf-eater, eating only the foliage and flowers of a few species of eucalyptus gum tree. To cope with this fibrous diet it has developed a set of efficient grinding teeth, and its digestive organs include an enlarged cecum—a part of its intestine—like that of the rabbit. The cecum contains bacteria that work on the food to break down the cellulose.

Greater gliders sleep among dense vegetation or in hollow trees by day and emerge at night to gather food. A glider will often travel some distance to find the right type of tree on which to feed. The female's pouch contains two teats, but only one offspring is born at a time. It crawls into the pouch and attaches itself to a teat, at which it is suckled for about four months before it emerges. Once out of the pouch, it clings to its mother's back and continues to be suckled at regular intervals.

Greater gliders have not suffered from the introduction of predators to the same extent as many other Australian animals. Foxes are their principal enemies, but the effect of fox predation on their numbers is minimal. Since the gliders are safe in the tall trees, and they rarely come down to ground level, few are taken. When they glide from one tree to another they usually aim to land near the top of the tree. Out of reach of most predators, they are also ignored by human hunters: their coats consist of long, sparse hairs, and are of no interest to the fur trade.

Ringtails

The common ringtail possum belongs to the same family as the greater glider, but it does not have a gliding membrane. It is one of 15 similar species of ringtail possums, all named for the distinctive ring-like curls at the tips of their tails. The common ringtail is about 16 in. long and inhabits the woods of eastern Australia, New Guinea and Tasmania, where it lives in small groups of up to three animals.

Spending its whole life in the trees, the common ringtail is a superbly adapted climber. Two toes on each foot are opposed to the other three toes giving the animal a good grip on the branches, and a muscular, prehensile tail effectively provides a strong fifth limb. It moves slowly from branch to branch in search of leaves, fruit and flowers, together with insects and other small invertebrates. The pouch opens to the front and conceals four teats, although the female gives birth to only two offspring at a time.

The honey possum

Many tree-dwelling marsupials include nectar and pollen in their diet, but the honey possum feeds on nothing else. It has been called "the hummingbird of the marsupials," an appropriate name in view of its diet and size—it is one of the smallest of the

ABOVE A ringtail possum demonstrates the curled tail tip from which it takes its name. Like many other tree-dwelling marsupials, these possums rely on their strong, muscular tails to provide support and security as they climb among the branches foraging for food. The tails of many species are wholly or partly naked to give a better grip on smooth bark.

marsupials. A large specimen measures 3.5 in. from its head to its rump, with another 4 in. of slender prehensile tail, and weighs up to 35 lbs. Females weigh on average one-third more than males.

The honey possum is only distantly related to the true possums, and is classified in a family of its own. It lives among the heathlands of southwest Australia, where the density and variety of flowering trees and shrubs are such that it can find flowers to feed from all year round. Adapted for the purpose, it has an elongated snout, a long, brush-tipped tongue and a bristly palate. The brushes scrape the nectar and pollen off its tongue as it is withdrawn from the flower. While feeding, it clings to the foliage with its tail and narrow, clawed feet.

Honey possums feed at night, sheltering during the day in nests among the foliage—often taking over abandoned birds' nests. They live alone or in pairs, and occupy overlapping feeding ranges of about 2.5 acres in size. The females are more aggressive than the males, and will chase the males out of their

MARSUPIALS

territory when they have young to feed. Each female produces up to four offspring, which stay in the pouch for some eight weeks. After that period they are left in the nest while the mother forages for food. She returns at intervals to suckle them until they become independent at about 11 weeks old.

The sleepy koala

One of the most familiar of all marsupials, the koala has endeared itself to generations of children thanks to its soft, furry, teddy-bear appearance. It is sometimes called the "native bear" or "koala bear," although it is not in fact a bear at all. The first white Australian colonists, however, were not sentimental about the animal, and shot millions of the creatures for their dense, ash-gray fur. Two million skins had been exported from Australia by 1924. As a result, the koala appeared to be heading for extinction by the 1930s, and only strict protection allowed the population in some parts of its range to recover to a healthy level. Paradoxically, some areas are now overpopulated by koalas. In small, isolated forests it is easy for the animals to eat all their food supply, and on several occasions koalas have been removed from such areas by conservationists and released in places where food is more abundant.

Overpopulation of an area of forest occurs because the koala is extremely fussy about what it eats. It is a browser, and if necessary it will feed on a variety of leaves, but prefers the foliage of only a few species of eucalyptus gum trees. Koalas will congregate wherever these trees grow, and they feed on them exclusively until the supply runs out. In the past, when vast tracts of continuous eucalyptus forest stretched throughout eastern Australia, supply was not a problem. The koalas would simply move on when one area was exhausted of food. Today, when clearance and uncontrolled fires have made the forests both smaller and more scattered, they do not have the same freedom of movement.

The concentration of koalas in fragmented habitats may have contributed to the spread of a serious bacterial infection that has recently reached epidemic proportions within the koala population. The bacteria produce a range of illnesses in koalas, often leaving them too weak to fend for themselves. The same bacteria infect a host of other mammals, and even cause some illnesses in humans.

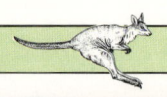

ABOVE Their long tails uncurled, a pair of eastern ringtail possums peer about in the unfamiliar daylight. Normally active by night, ringtails spend their time feeding high in the rain forest trees of tropical Queensland and New Guinea. They move slowly and quietly from branch to branch in search of leaves, fruit and flowers, together with insects and other small invertebrates such as spiders. By day they usually retire to sleep amid thick vegetation, or in leaf-lined nests built within hollow trees.

MARSUPIALS

Pouched heavyweights

The wombats are burly, ground-dwelling marsupials with a superficial resemblance to the marmots of alpine Europe. The resemblance extends to their habits—for they are great burrowers—and to their dental structure. Like marmots (and all other rodents), they have no canine teeth, but they do have large, strong incisors that grow continually to make up for a high rate of wear. They also chew with characteristic sideways movements similar to the chewing actions of marmots.

Wombats have thick-set, heavy bodies, large heads, short muzzles, small eyes and short ears. Their short, powerful limbs are well adapted for burrowing. All the toes on their forelimbs, and four on each of their hind limbs, are equipped with powerful claws for digging. They have poor eyesight but good smell and hearing—like most animals that live underground.

MARSUPIAL CLASSIFICATION: 8

Tarsipedidae

The honey possum, *Tarsipes rostratus* is the only member of the family Tarsipedidae. This nectar-drinking marsupial is restricted to the heaths of southwestern Australia.

Phascolarctidae

The koala, *Phascolarctos cinereus*, is also the sole representative of its family, the Phascolarctidae. It feeds almost exclusively on eucalyptus leaves, and inhabits the eucalyptus forests of eastern Australia.

Vombatidae

There are three species of wombat in the family Vombatidae. The common wombat, *Vombatus ursinus*, is found in forest and heathland in southeastern Australia. The southern hairy-nosed wombat, *Lasiorhinus latifrons*, lives on grasslands and among the sparse forests of central southern Australia, while the much rarer northern hairy-nosed wombat, *L. krefftii*, is limited to a single site in eastern Queensland.

TOP AND ABOVE Though their closest relatives are the tree-dwelling koalas, wombats are superb burrowers. The burrow acts as protection for the wombat. When a predator tries to enter one of its tunnels, the wombat will turn round and block the entry with its rump. Wombats occupy the same ecological niche as rabbits do in Europe.

Wombats are active by night and spend the daytime in their burrows sheltering from the fierce Australian sunshine. The burrows are up to 20 ft. wide and may stretch for 98 ft. in length. They have several entrances and a number of inner chambers that keep cool during the day but are warm during winter nights.

Common wombats live in eucalyptus forests and grasslands from the south of Queensland along the mountains of the Great Dividing Range to Victoria and South Australia. Two subspecies occur in Tasmania,

MARSUPIALS

The 60 or so species of kangaroos and wallabies are confined to Australia and New Guinea.

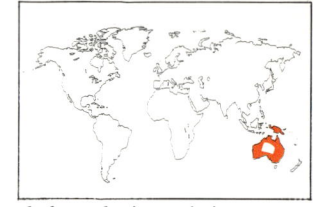

Wombats are social animals but their activity, even during mating, is kept to a minimum to conserve water and energy. Males tend to avoid one another as much as possible during courtship but, if a confrontation is unavoidable, one animal will try to bite the flank of the other. Its opponent kicks out with its powerful hind legs in self-defense.

There are two other species of wombat: the southern hairy-nosed wombat and the northern hairy-nosed wombat. Both species are slightly smaller than the common wombat, but they are similar in appearance. The hairy-nosed wombats live in drier areas and are efficient at conserving body water. Their urine is concentrated, and the water content of their feces is kept to a minimum. Like the common wombat, they are further able to cut down water loss during the day by reducing their rate of breathing while remaining inactive in their burrows. Hairy-nosed wombats are solitary, and only mix with others during the mating season. Only one population of northern hairy-nosed wombat survives, living in arid woodlands in Queensland.

Kangaroos and wallabies

The two families that make up the kangaroos and wallabies include about 60 species of marsupials. Together, their natural ranges extend through most of Australia, Tasmania, New Guinea, and some nearby islands. Some species are very large, while others are the size of rats. They all have small heads compared to their total body size and, while their forelimbs are poorly developed (except in tree kangaroos), their hind limbs are very powerful and are built for hopping.

The forelimbs are not used for locomotion—the hind limbs provide all the power the animals need in order to move about. The large hind feet have two enlarged toes with long claws. Two of the remaining toes are bound in a common sheath, with only their claws separated. All in all, a kangaroo's body looks out of proportion, with the back half being so large and powerful in comparison to the front. The front of the body has to be light to give the correct weight distribution for efficient hopping. The tail is long and strong—it serves as a counterweight when the animals are hopping and as a prop or a "seat" when they are grazing. Kangaroos can hop swiftly, but stamina is more important than sprinting speed in the open, bushy countryside of Australia where cover is sparse.

ABOVE The brush-tailed rat kangaroo (or brush-tailed bettong) is a very rare species. It has a long tail ending in a tuft of hairs. Rat kangaroos are thought by zoologists to be the ancestors of all kangaroos.

where they are still numerous, and in Flinders Island in the Bass Strait, separating Tasmania from the mainland. Many populations have been destroyed by farmers who regard the wombats as vermin. They have also suffered through competition from rabbits, brought to Australia by European settlers over the last two centuries.

Common wombats reach about three feet in length—both sexes are roughly the same size. They have short, pointed ears, cheek pouches, a stump of a tail and shovel-like paws. They eat mainly grasses and roots (though some also eat fungi), and have teeth that are adapted to cope with a fibrous diet. The incisor teeth are perfect for gnawing through tough grasses and, like those of rodents, they never stop growing. Females have pouches containing two teats for suckling their single young. Like so many marsupials, they provide a fine example of convergent evolution—the process whereby completely unrelated animals evolve similar adaptations to a certain environment and life-style, and consequently come to resemble each other.

MARSUPIALS

Australian "antelopes"

Kangaroos and wallabies are herbivores. In many ways, they are the Australian equivalent of the antelopes, adapted for browsing on tough plants and small bushes in the arid interior of Australia. Like ruminant placental mammals, such as cattle and sheep, they have bacteria in their gut to aid digestion. The bacteria cause the food to ferment, breaking it down slowly and releasing its energy over a long period of time—this is vital if the animals are to survive prolonged droughts or food shortages caused by fires.

Some species, such as red kangaroos, breed at any time of the year, though they are unlikely to breed at all during droughts. Others, like gray kangaroos, are usually seasonal breeders, producing offspring during the summer months. The Tasmanian wallaby, in contrast, has a very restricted breeding season—most of the young are born in late January.

Heading for the pouch

The smaller wallabies weigh less than 0.02 oz. at birth, while kangaroos, such as the red or the gray, weigh about 0.04 oz. Gestation lasts around one month. The newborn animals clamber up the belly of the mother, over the lip of the front-opening pouch, and attach themselves to one of the teats inside. They stay there for between 5 and 12 months (depending on the species), and suckle from the same teat for 2 to 6 months after they leave the pouch. They stay close to their mother until they are sexually mature—this may be at 18 months in larger species. In most species only one offspring is born at a time and twin births are rare. However, in the musky rat kangaroo, twins are the norm. In almost all species the female may mate shortly after giving birth. A dormant embryo results, and does not develop further until the offspring of the previous mating leaves the pouch.

When the Aboriginal people of Australia hunted kangaroos and wallabies for their skins and meat, they had little effect on the animals' populations. But the arrival of European settlers soon upset the ecological balance in the continent. They introduced rabbits, which multiplied quickly and ate the grasses and other plants that kangaroos and wallabies had relied on. They also introduced foxes, which preyed on smaller wallabies, and brought cattle, sheep and goats to graze on the land. Large numbers of kangaroos were shot because they ate the food of the livestock.

ABOVE Rock wallabies are agile animals that live in the rocky regions of Australia. With their long tail, used for balancing, and the non-skid pads on their hind feet, they are able to leap chasms over 13 feet wide and climb sheer rock faces.

BELOW Three methods of locomotion of the potoroo rat kangaroos: the normal four-footed walk (A); a short hop with tail hanging down—not used as a counterbalance (B); and the hopping gait with the tail lifted as a counterbalance (C).

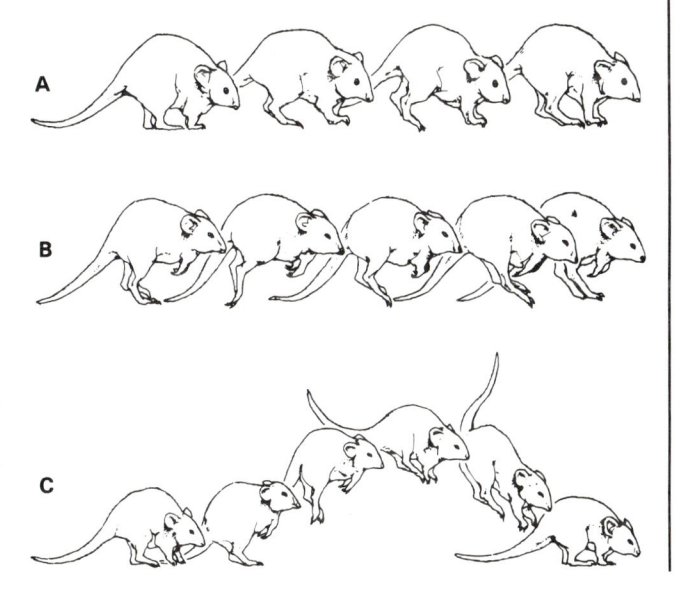

635

MARSUPIALS

ABOVE A rock wallaby searches around a tree stump for leaves and twigs to make a nest. Rock wallabies face competition from goats that take shelter in caves or in the shade of large rocks, forcing the native rock wallabies into the punishing heat of the day.
BELOW The first stages of aggression between two male quokkas, one of the smallest wallabies. They sniff each other before fighting (top); and exchange blows with their forepaws (bottom).

Rat kangaroos

The rat kangaroos range in size from some 16 to 35 in. long, including their tails. They bear a superficial resemblance to rats, but their hind legs are longer than their forelegs and they have much shorter tails in proportion to their body lengths. They inhabit a range of environments from deserts to rain forests, but most live on lightly wooded grasslands. Rat kangaroos are mainly nocturnal, and rest by day in nests located on the ground near tufts of grass and bushes. They often use their prehensile tails to carry grass, timber, and roots used for nest construction.

Tree kangaroos

Tree kangaroos live in the mountain forests of Queensland and New Guinea, at heights of up to 9800 ft. As their name suggests, they are tree dwellers, and they can leap nimbly from branch to branch covering 30 feet or more in one jump. Their forelegs are large and strong, unlike those of other kangaroos, and their forefeet are almost as large as their hind feet. They have separate toes with curved claws and rough soles for gripping the branches. Their tails are long and act as supports when the animals are climbing. They have soft hair, longer on the nape of the neck. Tree kangaroos have black faces, grayish brown backs, and yellowish white bellies.

Tree kangaroos are skilled climbers and can leap several yards from a higher branch to a lower one. At night they often come out of the trees and feed on the ground in the relative safety of darkness. They are able to jump down to the ground from great heights—up to 59 ft. or more—without injury. On the ground they hop like other kangaroos, but they need to bend further forward to counterbalance their long tails.

They feed on leaves and fruit that other kangaroos are unable to reach and live in small groups made up of one male and several females. The populations of all seven species are thought to have declined, but, since they live in inaccessible areas, detailed information on their numbers is scarce. The main threats come from loss of habitat because of deforestation.

Hare wallabies

Hare wallabies are some of the smallest of the kangaroos and wallabies, the largest of them weighing a little less than nine pounds. They are called hare wallabies because they are about the same size as a hare, can move swiftly, and can make long, high leaps. Mainly nocturnal, they feed at night and rest during the daylight hours in the shelter of long grasses or in lairs like those of hares.

Hare wallabies once lived over much of the arid and semi-arid interior of Australia, but their numbers have recently decreased and their populations have become fragmented. In many places, the decline of the hare wallabies coincided with the enforced removal of the Aborigines from their homelands. The Aborigines would burn fires in winter to clear the land for hunting. By removing thick vegetation, a variety of new plants were able to grow, improving the food supply for the hare wallabies. Once the Aborigines left, the variety of plants dwindled, dead scrub built up and summer fires devastated large areas and the wallabies that lived in them.

Rock wallabies

The rock wallabies are so called because they live in the rockier areas of Australia. The short claws and tough pads on their hind feet give a good grip on bare surfaces, and they can leap over 13 feet across fissures. They shelter from the heat in the shade provided by rocks, and choose the most inaccessible spots for the safe rearing of their young.

Some rock wallabies have striking colors. The yellow-footed rock wallaby has yellow and brown rings on its tail, while other species develop purple areas on their backs at certain times of the year. The animals used to be hunted for their pelts, but now the main threat comes from rabbits and goats, which compete for the same food supply.

"Organ grinders"

The nail-tail wallabies are unusual in that they have a horny tip to the tail, rather like a human fingernail in appearance. Nail-tail wallabies are also known by local Australians as "organ grinders," from their habit of moving their arms in circles as they hop. They used to live in shrublands and forests from Queensland south to Victoria, but have been exterminated over most of their former range.

The few remaining populations are still threatened. The enemies of nail-tail wallabies include foxes and feral dogs (domesticated dogs that have returned to the wild), but the main reason for their continued decline is clearance of their shrub and forest habitats.

TOP A young tammar wallaby (or dama) will stay close to its mother for several months after leaving the pouch. Now very rare in the wild, many are kept in captivity where they breed well.
ABOVE Tammar wallabies are social mammals that live in small colonies. There are about 11 separate populations in southwest Australia.
PAGES 638-639 A male red kangaroo (right) bends down to investigate the pouch of a female that contains a large joey or young kangaroo.

637

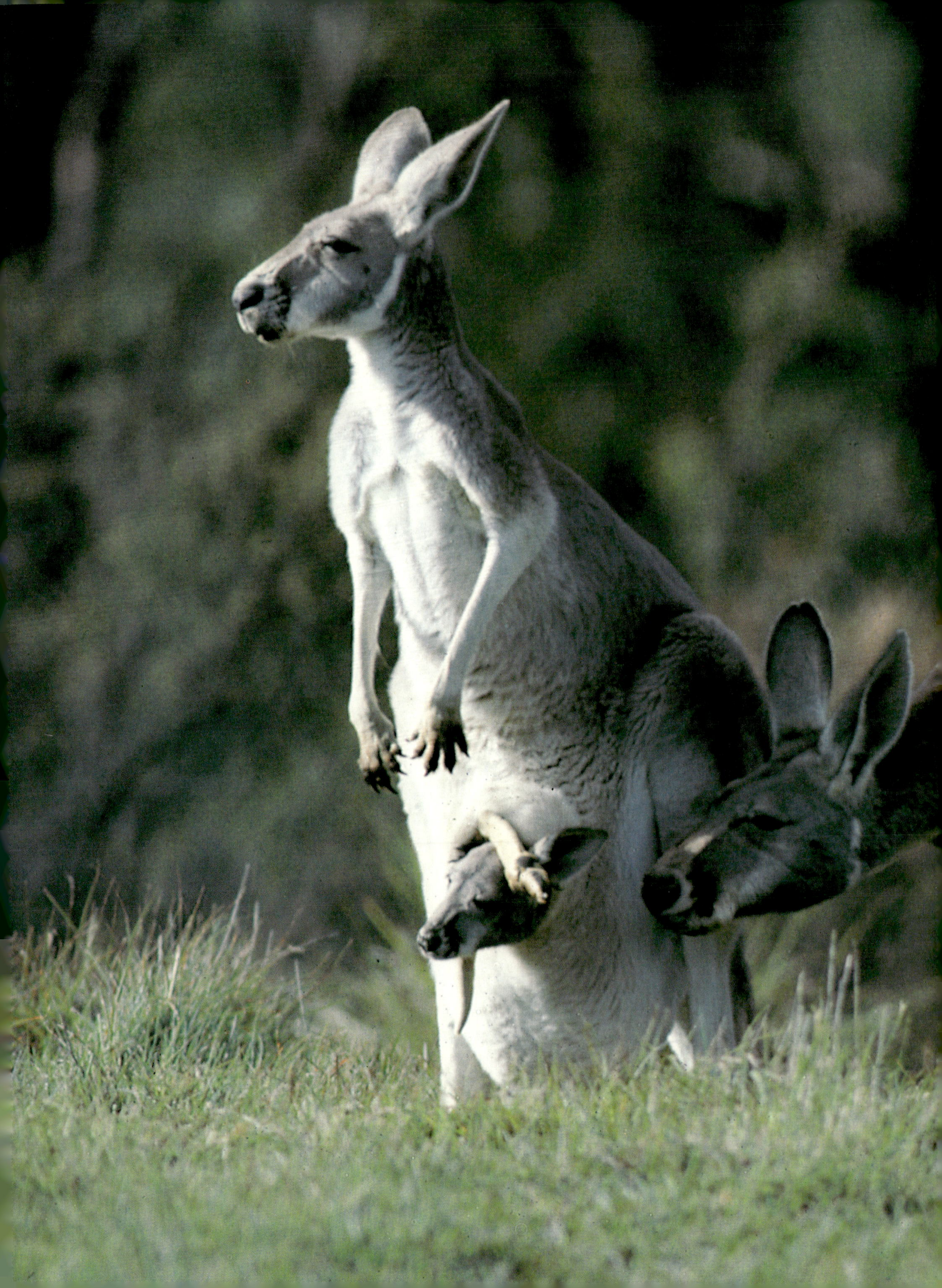

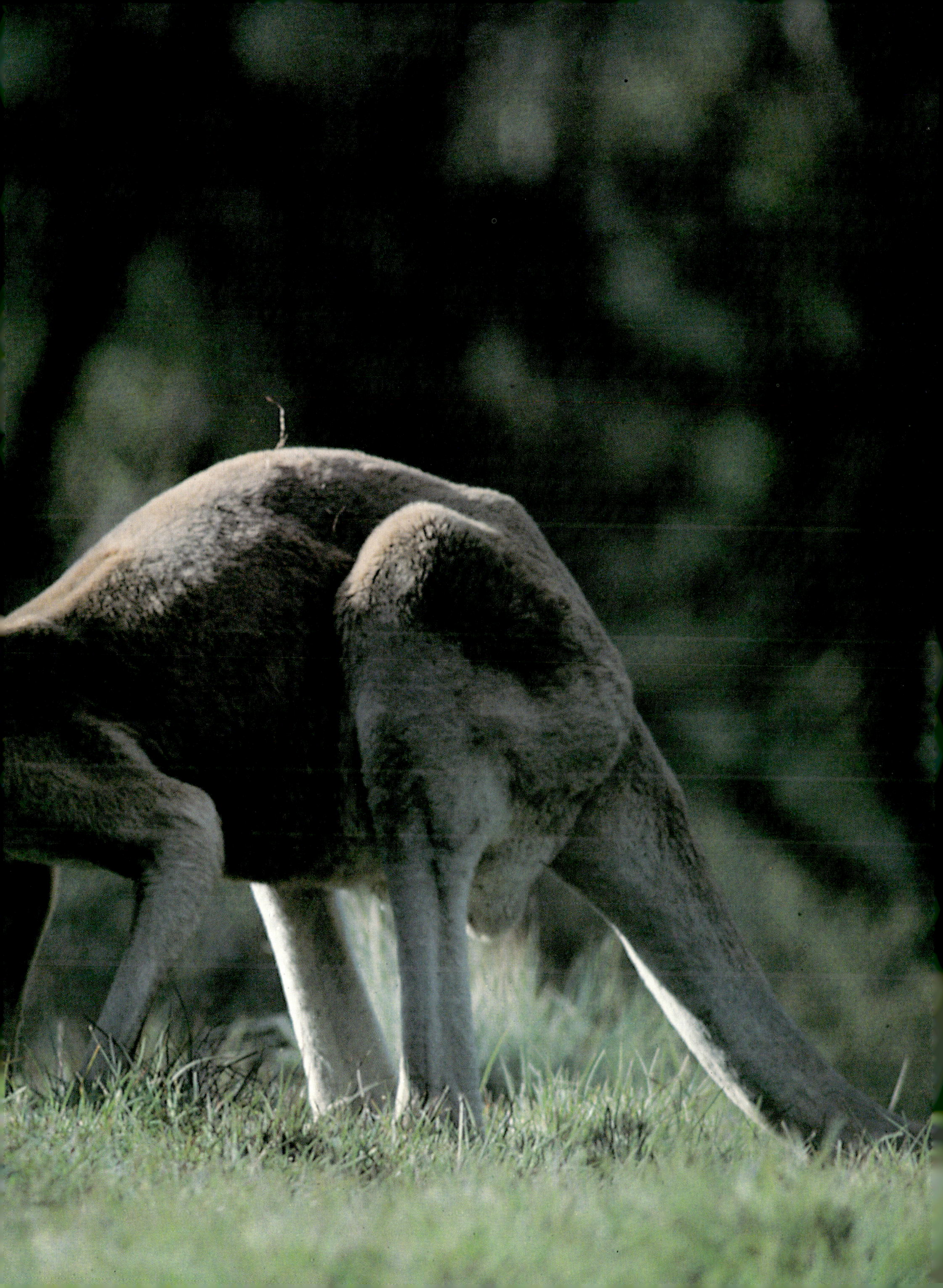

THE KANGAROOS
— SURVIVING IN THE OUTBACK —

Kangaroos are probably Australia's most well-known animals. In their appearance and in the way they move they are quite unlike any creatures found on other continents, and their great size sets them apart from the other Australian marsupials. Sociable animals, they normally live in small groups. Only rarely are they seen alone. Each group roams continually over a shared home range in search of food, especially relishing the lush pastures that spring up suddenly after a storm. At such a time, it is possible to see thousands of kangaroos feeding in the same area, although this does not constitute a true social gathering—merely a number of small groups sharing a rare feast.

Kangaroos can stand long periods without drinking, but during droughts they are known to dig holes in the ground to find water. At full hop they are fleet-footed animals, able to reach speeds of about 31 miles per hour when bounding away from dingoes. They cannot keep up such speeds for long, but if they are cornered by dingoes they can resort to self-defense—kicking out at their attackers with their powerful hind legs. When moving at speed, they avoid overheating by sweating copiously, in the same way that people sweat during heavy exercise to release the heat generated by their muscles. As soon as they stop moving they stop sweating; otherwise too much water would be lost from their bodies. In the heat of the Australian interior, the problems of water loss are all too acute. Apart from man, drought and bush fire in the dry terrain are the kangaroos' main enemies.

Slaughtered by the millions

It was the eastern gray kangaroos that suffered the most at first, after the arrival of the white man. When the land was turned over to cattle, sheep and goats, the kangaroos were shot because they were regarded as pests.

In the late 19th and early 20th centuries, many millions of kangaroos of all species were shot for bounty payments paid by the authorities that wanted the animals exterminated. Such slaughter was in addition to the

640

hunting carried out for meat and skins. Kangaroo meat increasingly became used for pet food, especially after the 1950s when improved methods of refrigeration led to a further increase in shooting. From 1954 to 1976, in the state of Queensland alone, a minimum of 200,000 and a maximum of over one million were killed annually. All these killings were legal. Under the severe pressure from hunters, kangaroo numbers fell alarmingly in many areas. Nevertheless, their populations are still very large in some areas—the state of New South Wales has over 2,000,000 red kangaroos.

Kangaroos are tough, adaptable animals, and since they have survived the onslaught of white settlers for the last hundred years, it is hoped they will continue to remain numerous in the future. Over Australia as a whole, the total number of gray and red kangaroos still probably exceeds the human population.

ABOVE LEFT A gray kangaroo uses its long, strong tail stretched out as a counterbalance when hopping. Kangaroos have been known to make leaps of 26 feet long and to jump as high as 11 ft. 6 in. They can also reach speeds of up to 30 miles per hour.
ABOVE RIGHT A young gray kangaroo, not yet weaned, suckles from a teat in its mother's pouch. It will suckle in this way for several months after leaving the pouch. Offspring stay close to their mothers until they are sexually mature—from about 18 months to three years old.
RIGHT A sequence of drawings showing how a joey climbs into its mother's pouch, even when the youngster has grown quite large: it first inserts its head and forearms into the pouch (top), then performs a somersault inside (center), so that only its head, hind legs and tip of the tail protrude (bottom).
FAR LEFT A group of gray kangaroos rest during the heat of the day. Kangaroos are becoming less nocturnal and more sociable in habit since settlers cleared the forests, forcing kangaroos out into the open.

MARSUPIALS

MARSUPIAL CLASSIFICATION: 9

Potoroidae

The rat kangaroos belong to the family Potoroidae. The family comprises the four species of bettongs or short-nosed rat kangaroos (genus *Bettongia*); the three species of potoroos (genus *Potorous*); the musky rat kangaroo, *Hypsiprymnodon moschatus*; the plains or desert rat kangaroo, *Caloprymnus campestris*; and the rufous rat kangaroo, *Aepyprymnus rufescens*.

Macropodidae

All other kangaroos and wallabies belong to the family Macropodidae, which comprises 11 genera. They include the seven species of tree kangaroos (genus *Dendrolagus*); five species of hare wallabies (genera *Lagostrophus* and *Lagorchestes*); two species of nail-tail wallabies (genus *Onychogalea*); the four species of pademelons or scrub wallabies (genus *Thylogale*); ten species of rock wallabies (genus *Petrogale*); the quokka, *Setonix brachyurus*; the swamp or black wallaby, *Wallabia bicolor*; and the five species of forest wallabies (genera *Dorcopsis* and *Dorcopsulus*). The remaining 14 species belong to the genus *Macropus*, including the Tammar or Dama wallaby, *M. eugenii*; the wallaroo or hill kangaroo, *M. robustus*; the eastern gray kangaroo, *M. giganteus*; and the red kangaroo, *M. rufus*.

Forest wallabies

The forest wallabies of New Guinea are not as well adapted for jumping as other wallabies—probably because jumping is not the best way of moving about in the dense, damp rain forests where they live. Their coloring is mainly dark brown and provides good camouflage in the shade of the rain forest cover. Native people eat the animals, which are also caught by their dogs, but the greatest threat by far to forest wallabies is tree felling for the export of timber.

Pademelons live in thick scrubland, forest, savannah, and dense grassland. They are nocturnal wallabies and live in small groups. Though much hunted, they still survive in eastern Queensland, eastern New South Wales, Tasmania and New Guinea.

The quokka is one of the smallest wallabies and is

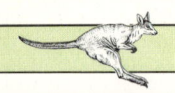

TOP Living in some of the hottest parts of Australia, red kangaroos conserve their energy by being inactive during the day.
ABOVE Two male gray kangaroos fight for possession of a female or group of females. They interlock forearms and try to push the opponent to the ground, using their tails for support.
FAR RIGHT Although kangaroos are supposed to feed at night, they will—like these two gray kangaroos—venture into a shady, open eucalyptus forest to browse.

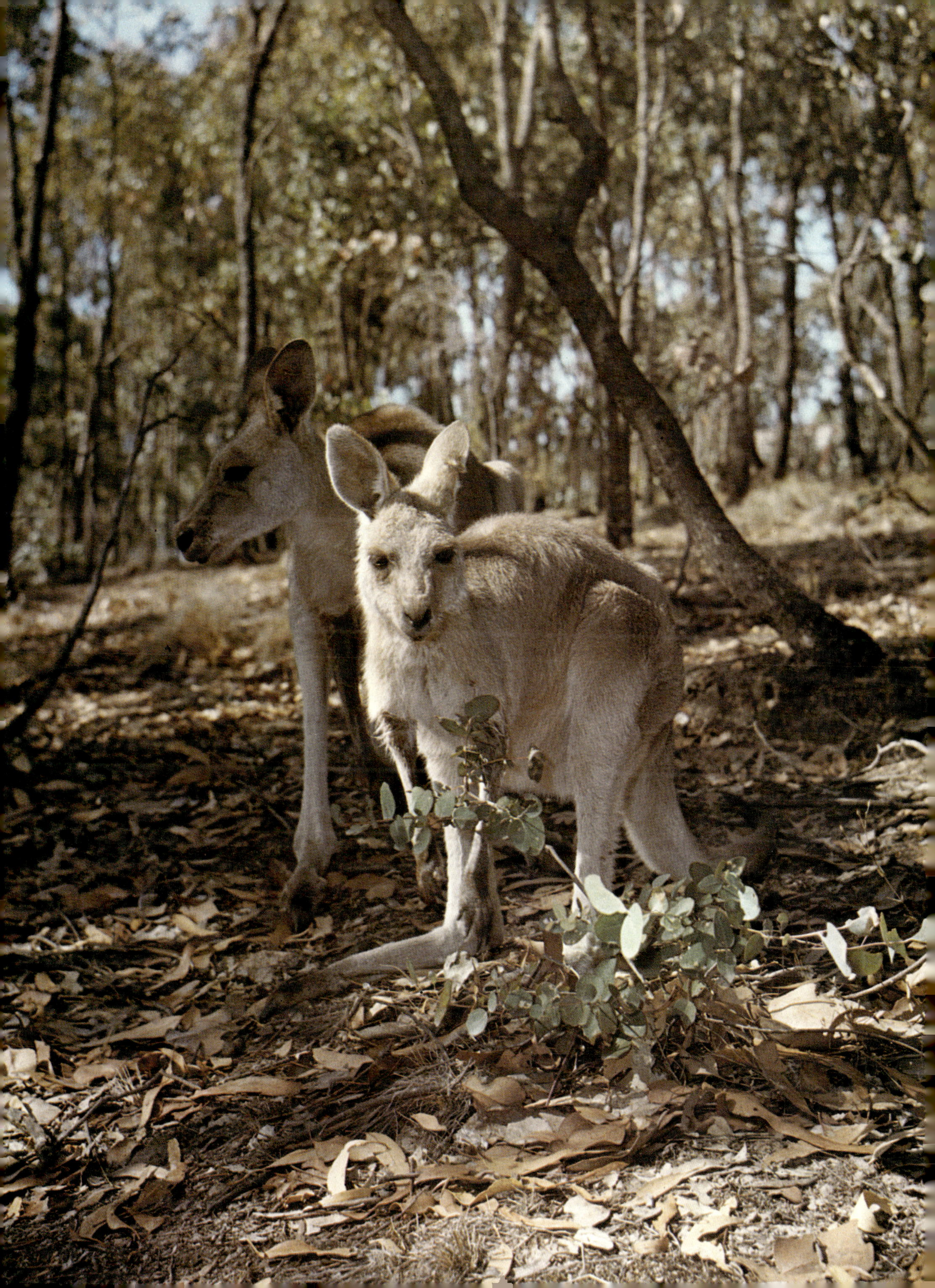

MARSUPIALS

ABOVE Female gray kangaroos carry their young in the pouch for about six months before the joey is able to stand on its own. When the mother has another baby, the firstborn continues to follow her around to suckle. BELOW Three aspects of a kangaroo's behavior: a minute and undeveloped kangaroo, having found its way into the mother's pouch immediately after birth, firmly attaches itself to a teat (A); a male kangaroo (right) courts a female before mating (B); a kangaroo licks its forepaws to keep cool during a hot day (C).

unusual because its weak tail cannot be used as a support. It lives in thickly vegetated areas, especially swampy thickets, where it can tunnel through the undergrowth. Its fur is short and thick, and the animal weighs from 6.5 to 9 pounds. The quokka used to be common throughout southwest Australia, but there are now only a few areas where it survives without being disturbed. Two of its remaining strongholds are Rottnest and Bald islands off the west coast of Australia.

The giant kangaroos

The largest living marsupials are the red kangaroos and the gray kangaroos, which can measure over 8 ft. from head to tail and weigh as much as 198 lbs. There are two species of gray kangaroo: the eastern gray lives mainly to the east of the Great Dividing Range from Queensland through New South Wales to Victoria; the western gray lives to the west of the mountains from the southern part of Western Australia across to western Victoria. The Great Dividing Range forms a barrier that prevents the two species from interbreeding. Though the eastern gray kangaroo is gray in color as its name suggests, the western gray kangaroo is mainly brown.

The red kangaroo has the widest distribution, occurring over most of Australia west of the Great Dividing Range, except for the extreme north, east, and southwest. It is the largest of the species, and continues to grow throughout its life. Animals kept in captivity have lived for 28 years, but those in the wild are more likely to live for up to 22 years.

Coat colors

There is little difference in coat color between male and female gray kangaroos, but males are much larger in size. Red kangaroos, by contrast, do show color differences between the sexes; males are generally a deep rusty-red and females are bluish gray. Male reds are also considerably larger than females—in some cases up to twice as large.

Male red kangaroos fight for mating rights when females are in heat. Before they engage in battle they perform a ritual "four-legged" walk, then stand upright for ritual grooming and scratching. When they finally confront each other, they interlock forearms and try to push one another to the ground. The winner will immediately chase his rival away.

INSECTIVORES AND ELEPHANT SHREWS

THE INSECT HUNTERS

Running, burrowing and swimming to catch their prey, insectivores have a ferocity and an appetite to make up for their small size, some tackling creatures far larger than themselves

INSECTIVORES AND ELEPHANT SHREWS

Common tenrec

Desert hedgehog

European common shrew

Short-tailed shrew

North African elephant shrew

Star-nosed mole

Giant otter shrew

European mole

646

INSECTIVORES AND ELEPHANT SHREWS

At the end of the Cretaceous period, about 65 million years ago, as the Earth neared a crisis that led to the extinction of the dinosaurs, a small group of placental mammals emerged. These were the insectivores—forerunners of today's moles, shrews, and hedgehogs. They were tiny insect-eating creatures (three million of the smallest species of modern-day shrew probably equals the weight of the seven-ton dinosaur *Tyrannosaurus rex*), constantly in search of food to fuel their active bodies. It is now believed that all placental mammals—including the primates—descend from these first insectivores.

There are more than 340 species of insectivores. They have a highly developed sense of smell, generally long snouts and small brains, with little folding to increase their surface area. They are flat-footed, walking on the soles and heels of their feet rather than on their toes. Their testes do not descend into a scrotal sac, and they have a cloaca—a passage into which both the urine and feces empty. Marsupials also have a cloaca but, unlike marsupials, female insectivores have

ABOVE RIGHT The forelimbs of a European mole are adapted for digging: the five claws are large and the forefeet are turned outward. It has no external ears and is almost totally blind.

BELOW The map shows the geographical distribution of some of the insectivores and the elephant shrews.

PAGE 645 The European hedgehog is a good swimmer and will cross streams in search of food.

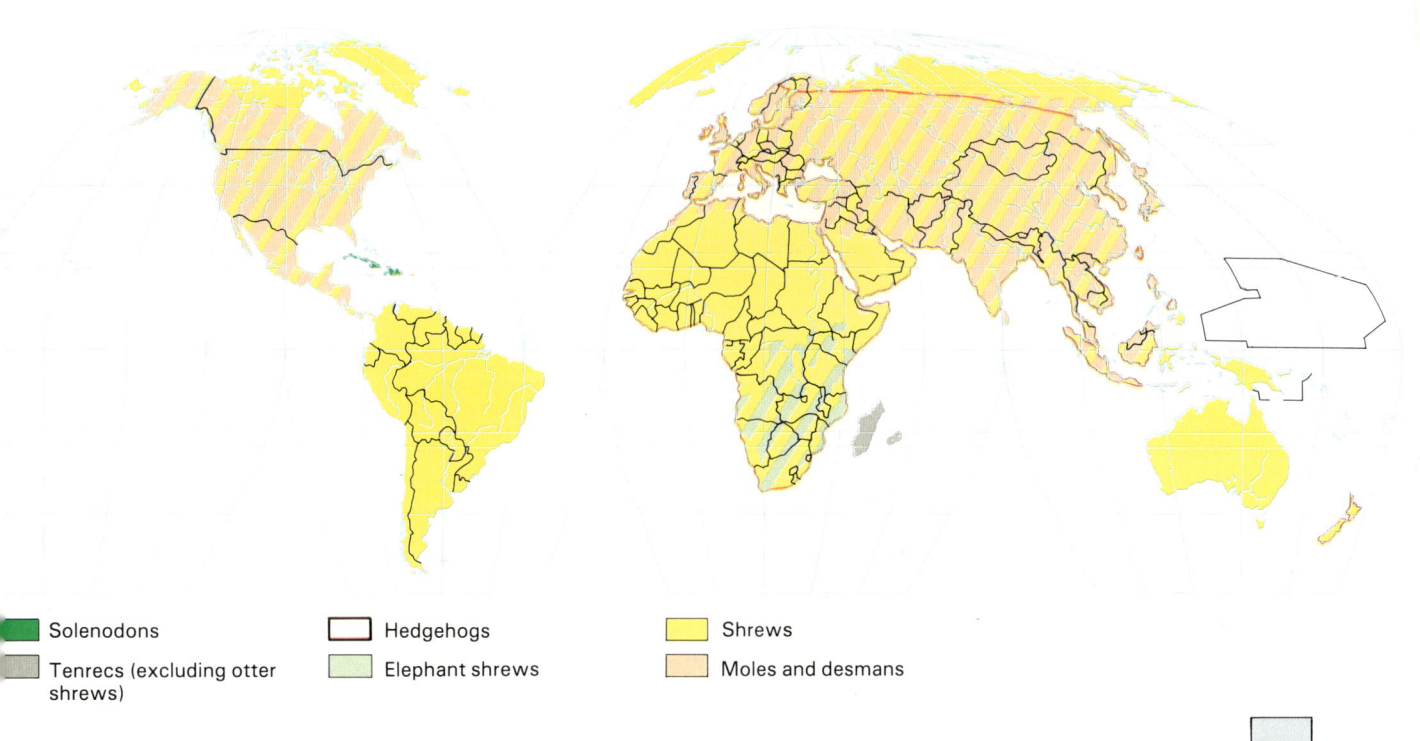

- Solenodons
- Tenrecs (excluding otter shrews)
- Hedgehogs
- Elephant shrews
- Shrews
- Moles and desmans

647

INSECTIVORES AND ELEPHANT SHREWS

a placenta that enables the embryo to develop fully inside the womb.

There are four main insectivore shapes. The shrew-like body type is slender, with short legs and a relatively long tail. The spiny type—the hedgehogs and related tenrecs—has a stout, strong body and a short tail, and is protected by a dense covering of spines on its back. The otter-like body shape belongs to the aquatic insectivores, known as the otter shrews. They have a long slender body covered with soft fur, webbed feet and a flattened tail to propel the animal through water. The mole-like body shape is typical of animals that live underground—the moles and related desmans. These have robust bodies, muscular forelimbs with large claws for burrowing, and a stump of a tail. Most insectivores have small eyes and ears that may be more or less useless in some cases. Insectivores generally have short legs.

Widespread habitats

Insectivores live in a wide variety of habitats. Some species, like the solenodons of Haiti and Cuba, are extinct or on the verge of extinction. Others, especially moles and shrews, have succeeded in colonizing large areas of the world. But there are also large areas on the world map where there are no insectivores at all. There are none in Australia and New Zealand, none in most of South America, and none in the polar regions, which proved too cold for small animals unable to retain enough body heat. In some cases, families have evolved in isolation: solenodons in the Caribbean, tenrecs in Madagascar and the nearby Comoro Islands, and otter shrews (also in the tenrec family) in Central and West Africa.

Although insectivores live successfully in water (otter shrews) and underground (moles), most confine themselves to the humid environment of woodland undergrowth. They eat mainly insects and invertebrates such as earthworms. The rate at which body tissues are broken down and built up again (metabolic rate) increases in mammals as body size decreases. Most insectivores are very small and therefore need to eat large amounts of nutritious, energy-rich food. The tiniest mammal in the world is an insectivore—the pygmy shrew may be as small as 1.5 in. long and weigh less than 0.07 oz. The largest of the insectivores is the common tenrec, which grows up to 16 in. in length and weighs about 3 lbs. 5 oz.

TOP **Hedgehogs** are agile and fast-moving animals. They can climb well, using their long legs (normally hidden) and strong claws to gain a hold on walls or fences. They are often trapped at the bottom of cattle grids or killed while crossing roads.

ABOVE **The pygmy white-toothed shrew**, an inhabitant of Africa, is the smallest mammal in the world. Measuring around 1.5 in. long and weighing less than 0.07 oz., it lives in areas of lush vegetation, forest, scrub, and savannah.

INSECTIVORES AND ELEPHANT SHREWS

ABOVE The water shrew can stay underwater for 20 seconds while it hunts water animals such as insects, leeches, toads and fish. The air trapped in the animal's coat gives it a silvery sheen and makes the shrew more buoyant in the water.

Protective spines

Small animals like insectivores need an efficient means of defense against predators. The most obvious defense is the spiny coat of the hedgehog (the spines are modified hair). But some of the Madagascan tenrecs are also covered with protective spines and, like the hedgehogs, are able to roll themselves into an almost impregnable ball. When the animal is alarmed, the muscles from which the spines grow contract, causing the spines to stand up. At the same time, the animal draws its paws tightly into the body and lowers its head onto its chest, rolling itself into a ball and making its soft, unprotected belly almost impossible for a predator to reach.

One of the best forms of defense is to burrow underground, placing a thick layer of hard earth between prey and predator. Moles are experts at doing this. The golden mole family contains the best burrowers of all. Their underground homes are complex in layout, reach depths of up to 3 ft. and contain up to 820 ft. of tunnels.

Adaptations to water

The ability to swim and hunt underwater has been achieved with varying degrees of success by several groups of insectivores. As well as containing the familiar burrowing mole, the mole family includes species that are aquatic (the desmans) or semi-aquatic (the star-nosed mole), and all species can swim well. The two species of desmans are highly adapted for swimming—their ears and eyes can be opened and closed by flaps of skin acting as valves, and they have webbed toes, waterproof fur and powerful hindlegs to propel them through the water.

Hedgehogs can also swim across streams in search of food or to escape from danger, but some of the tenrecs are the animals most well equipped for life in water. The aquatic tenrec and the three species of otter shrews look very similar to otters. They all have streamlined bodies and flattened heads and, like otters, their ears and eyes are above the water when they are swimming at the surface. They have long, sensitive whiskers for locating prey underwater, and the aquatic tenrec and Ruwenzori least otter shrew have webbed feet and long tails. Their tails are slightly flattened from side to side and are used both for steering and for extra propulsion. Though they closely

649

INSECTIVORES AND ELEPHANT SHREWS

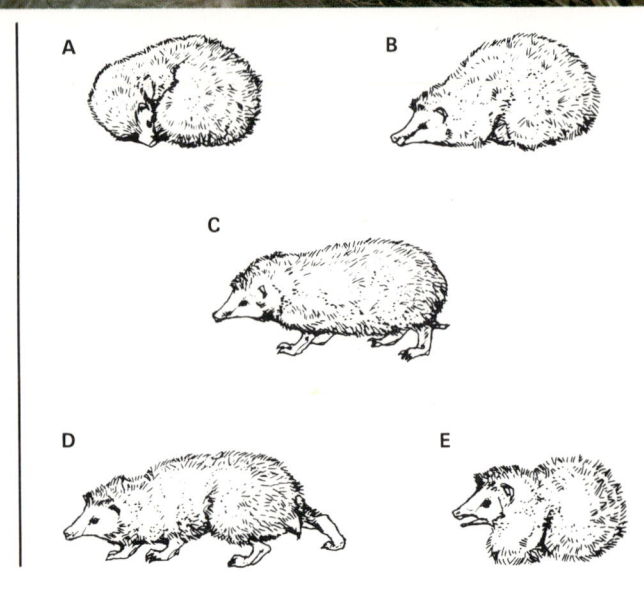

LEFT **The trunk-like snout of the Hispaniola solenodon is joined to the skull by a ball-and-socket joint; this gives the snout great flexibility in searching for food in holes and crevices, and for grappling with prey, such as insects, worms and small reptiles.**
CENTER LEFT **A European hedgehog spreads saliva over its spines, a habit that may attract a mate, get rid of parasites or drive off enemies.**
BELOW LEFT **The different postures of a hedgehog tenrec: rolled into a ball for maximum defense (A); protecting its soft belly only for partial defense (B); normal walking gait (C); running away from danger (D); aggressive defense with mouth open and spines erect (E).**

resemble each other in appearance, the aquatic tenrec and the otter shrews evolved in isolation: the former in Madagascar, the latter in mainland Africa.

Despite their adaptations, aquatic insectivores do not spend all their lives in water, for they also forage on land. Apart from the burrowing moles, all other species forage at ground level for some, or most, of their lives.

Solenodons

There are only two species of solenodons and both are on the brink of extinction. They live only in the Caribbean, in Cuba, Haiti and the Dominican Republic. Unless the governments of these countries provide strong protection and effective management for the forest reserves in which they live, these strange creatures that resemble giant shrews may soon disappear forever.

Solenodons are dying out mainly because of loss of habitat and predation—chiefly from introduced mongooses and feral cats in Cuba, and introduced dogs on the island of Hispaniola (Haiti and the Dominican Republic). They are nocturnal insectivores, spending the day in hollows in the ground or inside hollow trees. At night they feed mainly on beetles, insect larvae, crickets, millipedes, termites and earthworms, immobilizing their prey with poisonous saliva. They sometimes eat frogs and small reptiles, and perhaps even small birds. They rely on keen senses of smell and hearing, and use their long, flexible snouts to sniff out prey in cracks in the ground and in trees. They also emit high-pitched chirps, possibly as a means of echolocation—using the pattern of reflected sound waves to detect the presence of moving creatures.

INSECTIVORES AND ELEPHANT SHREWS

The tenrecs are native to Madagascar but the otter shrews occur in West and Central Africa.

INSECTIVORES CLASSIFICATION: 1

The order Insectivora, the insectivores, is one of the larger mammalian orders, containing a total of some 345 species ranging throughout Eurasia, Africa, North America, and the northern part of South America. They are divided into six families: the solenodons, the Solenodontidae; the tenrecs and otter shrews, the Tenrecidae; the golden moles, the Chrysochloridae; the moonrats and hedgehogs, the Erinaceidae; the shrews, the Soricidae; and the moles and desmans, the Talpidae.

Solenodontidae

The family Solenodontidae contains just two species, both confined to islands in the Caribbean Sea. The Hispaniola solenodon, *Solenodon paradoxurus*, occurs only on the island of Hispaniola, which is shared by the two nations, Haiti and the Dominican Republic. The Cuban solenodon, *Solenodon cubanus*, lives only on Cuba.

Tenrecidae

The family Tenrecidae has 34 species divided into three subfamilies. Members of the subfamily Oryzoryctinae are confined to Madagascar. They include 20 species of long-tailed tenrecs (genus *Microgale*); three species of rice tenrecs (genus *Oryzoryctes*); and the aquatic tenrec, *Limnogale mergulus*. The subfamily Tenrecinae is also restricted to Madagascar, and includes the common tenrec, *Tenrec ecaudatus*; the greater hedgehog tenrec, *Setifer setosus*; and the lesser hedgehog tenrec, *Echinops telfairi*. Members of the third subfamily, the Potamogalinae, occur in West and Central Africa, including the giant otter shrew, *Potamogale velox*; and the Ruwenzori least otter shrew, *Micropotamogale ruwenzorii*.

The Hispaniola solenodon has a head-body length of about 12 in. and a tail of up to 10 in. long. Its coat is grayish brown with yellowish brown flanks. The Cuban solenodon is much the same size and shape as its relative, but it has a slightly shorter tail and longer, finer hairs on its coat. Both species weigh about two pounds.

TOP The greater hedgehog tenrec lives on the island of Madagascar, off the east coast of Africa. It looks very similar to a European hedgehog but has evolved in isolation.

ABOVE The lesser hedgehog tenrec is native to Madagascar and is thought to be, along with other tenrecs, among the first mammals to have colonized the island.

The tenrecs

Tenrecs have evolved a greater variety of forms, adapted to different habitats and life-styles, than any other family of insectivores. They have colonized all available habitats except arid land. Of the 34 species of tenrecs, 31 have evolved in geographical isolation on the island of Madagascar, off the east coast of Africa. The only species that do not live in Madagascar are the otter shrews of the wetlands of Central Africa. Otter shrews look quite different from most other members

651

INSECTIVORES AND ELEPHANT SHREWS

ABOVE **The streaked spiny coat of a young common tenrec helps camouflage it in its grassy habitat. The young animal loses its spines as it grows older, replacing them with long, rigid hairs. Common tenrecs can have litters of up to 30 or more young (more than any other mammal). The mother is consequently forced to forage in daylight to keep up her milk supply, even though this makes her vulnerable to attack. When angry or frightened, the animal makes low hissing sounds.**

of the tenrec family, but certain features reveal that they are closely related. These include a low, variable body temperature and a common opening (cloaca) for the excretory and reproductive passages.

Tenrecs are nocturnal animals with poorly developed vision. Instead of using eyesight to detect their prey, they have developed keen senses of smell and hearing, and emit high-pitched chirps and tweets that may be used as a crude form of echolocation. Some tenrecs have especially long tails and long hind legs. They climb trees and jump from branch to branch, using the tails to help them balance. Those that live solely on the ground have short legs, and all that remains of the tail is a stump. The rice tenrecs have velvety dark-brown fur and look and behave like moles. They live in wet areas, especially rice fields that are flooded for most of the year, and burrow into the mud walls surrounding the fields. The common tenrec and the greater and lesser hedgehog tenrecs have protective, spiny coats—like those of true hedgehogs.

Some species of tenrecs have large litters—as many as 32 in the common tenrec. The offspring are generally born in the wet season. The female common tenrec has up to 29 nipples with which to feed her young—more than any other mammal. To keep up with the hefty demand for milk from her young, she is forced to forage for food in daylight hours when the danger from predators is greater both for herself and for her litter. To provide protection, the young animals have a striped camouflage, which they lose as they mature.

Heavy and spiny

The common tenrec is the heaviest of the insectivore species, weighing 4 lbs. 4 oz. or more, and measuring 16 in. without the tail. It has a spiny coat and a mere stump of a tail. Like all insectivores, it walks on the heels and soles of its feet (it is flat-footed, or plantigrade) and has well-developed claws. Its body temperature can vary as much as 11 degrees in 24 hours. By resting during the daytime and allowing its temperature to drop, it saves energy otherwise used to

INSECTIVORES AND ELEPHANT SHREWS

maintain a higher and constant body temperature. During dry or cold seasons the animals hibernate.

Common tenrecs eat insects, earthworms, small reptiles and amphibians, and even take small birds and mammals. They also sometimes eat fruit. When foraging on the ground surface, they dig small holes to find prey and detect others with sweeps of their long, sensitive whiskers. While hedgehog tenrecs simply roll themselves into a defensive ball in the face of danger, common tenrecs are far from passive when threatened. They advance toward danger, mouths wide open to inflict vicious bites—adult males have 0.6-in.-long canine teeth—and they swing their heads from side to side to drive spines into the aggressor.

Young common tenrecs sound the alarm by rubbing together the spines on their backs to produce a surprisingly loud noise. The main form of communication between individuals is through scent marking. They either deposit their feces and rub their anal glands on the ground, or rub the secretions from their eye and neck glands onto undergrowth.

Otter shrews

The three species of otter shrews live near rivers and streams in central Africa, close to the Equator. They eat insects, small fish, frogs and crabs. The giant otter shrew is the most aquatic of the species and also the largest, reaching up to 14 in. in length, not including the tail that may be a further 11 in. long. Its snout is dark brown on top and lighter underneath. The feet have five toes, but these are not webbed, power in water being provided by the powerful flattened tail. When swimming, the giant otter shrew holds it feet against its body and uses its tail like an oar. Although it spends much of its time in water and can swim quickly and skillfully, it is not exclusively aquatic and also forages on the ground. It makes its home in burrows close to the water.

The Ruwenzori least otter shrew has webbed feet and only a slightly flattened tail. The Mount Nimba least otter shrew spends only limited time in water, having neither webbed feet nor a flattened tail.

Golden moles

Golden moles are burrowing animals that live in a wide range of habitats, from forests, grasslands, riverbanks and swampy areas, to mountains, deserts and semi-deserts. They live all over Africa south of the

TOP **The lesser hedgehog tenrec hunts for food—insects, worms, and other small invertebrates—by night, and shelters in a nest of leaves or under rocks during the day.**
ABOVE **Although the common tenrec has an unpleasant smell and is protected by law, it is hunted for food by native Madagascans, and the fatty meat is sold quite openly in the markets. But deforestation and the use of pesticides are probably far greater threats.**

653

INSECTIVORES AND ELEPHANT SHREWS

The golden moles of Africa are mainly distributed south of the Equator.

Sahara Desert. Although they have much in common with the other moles, they also have many differences and are classified in a family of their own.

The golden moles take their name from the metallic sheen on their coat. Coat color varies from one species to another. The Grant's desert golden mole, the smallest species, is yellowish gold in color. Other species are bluish, purplish, green or bronze. Golden moles have thick-set bodies, no visible tail, short, strong limbs, and thick, tough skin. Their eyes are covered with hairy skin, their ears are covered by fur, and their noses are protected by a thick layer of hard skin—all adaptations to life underground. Their strong forelimbs have large claws for efficient digging; they can excavate almost 820 ft. of tunnels in their burrows. Golden moles living in the desert do not dig deep burrows because the dry sand would cave in. Instead, they keep close to the surface, digging with a breast-stroke swimming motion through the sand, leaving ridged tracks behind them.

Golden moles eat earthworms, insects and insect larvae, slugs, snails and spiders. Desert golden moles also catch legless lizards. At night and during heavy rain golden moles may forage on the surface of the ground, but they normally feed just underneath. Foraging above ground is a dangerous activity because snakes, birds of prey, mongooses, jackals and genets all prey on the moles. Usually the golden moles drag the food they find to the safety of a deep burrow. If they eat prey just under the surface, they risk being caught themselves, since a jackal can easily detect movement near the surface of the ground and dig out the golden mole before it has time to escape.

All golden moles are territorial. Those living in arid areas where there is little food have much larger territories than those living in areas with lush vegetation and a rich food supply. There are 18 species of golden mole, some now quite rare, especially in cultivated areas. The giant golden mole is threatened with extinction—its population has become fragmented and greatly reduced by economic and agricultural development throughout its range.

The moonrats

Moonrats, or gymnures, belong to the same family as the hedgehogs. There are five species that all occur in China and Southeast Asia. They have well-developed eyes and ears, long snouts and short legs, and all

TOP **Two young giant otter shrews wait at the entrance of their burrow for their mother to return from hunting. Otter shrews are the most aquatic of all insectivores, but they do not spend their entire lives in water. They forage on the ground and dig burrows near watercourses. Although similar to otters in appearance, they are, in fact, insectivores, not carnivores.**
ABOVE **The otter shrew has a streamlined body and a long, flattened tail to propel the animal efficiently through water. The three species of otter shrews are members of the tenrec family.**

INSECTIVORES AND ELEPHANT SHREWS

INSECTIVORES CLASSIFICATION: 2

Chrysochloridae

The golden moles make up the family Chrysochloridae. They occur in East, Central and southern Africa, and are divided into seven genera. The 18 species include the giant golden mole, *Chrysospalax trevelyani*; the Hottentot golden mole, *Amblysomus hottentotus*; Grant's desert golden mole, *Eremitalpa granti*; and the yellow golden mole, *Calcochloris obtusirostris*.

Erinaceidae

The family Erinaceidae contains two subfamilies. The 12 species of hedgehogs belong to the subfamily Erinaceinae. They range over Africa and Eurasia, and include the European hedgehog, *Erinaceus europaeus*; the four-toed hedgehog, *E. albiventris*; the long-eared hedgehog, *Hemiechinus auritus*; and the desert or Ethiopian hedgehog, *Paraechinus aethiopicus*. The moonrats belong to the subfamily Echinosoricinae and occur in south China, Southeast Asia, Indonesia and the Philippines. There are five species, including the greater moonrat, *Echinosorex gymnurus*, the lesser moonrat, *Hylomys suillus*, and the shrew hedgehog, *Neotetracus sinensis*.

ABOVE A long-eared hedgehog and her young use their pointed snouts and strong claws for foraging in leaf litter. The ears of these hedgehogs are so large that they cover their eyes when folded forward. Long-eared hedgehogs live in a broad geographical band from Libya in North Africa to northwest India, China and Mongolia.

except the greater moonrat have soft coats and short tails. The greater moonrat has a coarse, black coat with white markings on the forehead, neck and shoulders, and a long tail like that of a rat. It is about the size of an adult rabbit and can weigh up to 3 lbs.

Nocturnal insectivores, they rest in safe places during daylight, in holes in trees and in the ground. Having no protective spines, they are more vulnerable to predators than hedgehogs and are far more shy. They live in forested areas that provide good cover, but are now becoming rare as the hardwood forests of Southeast Asia are cut down and the valuable timber exported for the construction of fine furniture and other luxury uses.

Greater moonrats are good swimmers, and their diet includes mollusks, crustaceans, worms, insects and sometimes small fish. They are solitary animals with a keen sense of smell that they use to detect prey. They are territorial and use scent from anal glands to communicate with other moonrats living nearby. Moonrats occurring in the tropics breed at any time of year and have two or three young, except the shrew hedgehog, which has as many as five in a litter. The offspring are blind at birth.

The hedgehogs

Hedgehogs live throughout Africa and Eurasia, as far north as the limit of deciduous woodland. The most well-known species is the European hedgehog. Other members of the family include the long-eared hedgehogs and the desert hedgehogs.

Hedgehogs have thick, spiny coats, and when alarmed they roll into a ball to protect themselves, with the sharp spines facing outward. They have short, powerful legs and are good diggers. Long-eared hedgehogs live in arid areas and dig burrows to shelter

INSECTIVORES AND ELEPHANT SHREWS

ABOVE A desert hedgehog—recognizable by its large ears and the central stripe on its forehead—chews a large spider it has just caught. The desert hedgehog lives in arid areas from North Africa to Iraq, and may spend the hottest months in an inactive state (called estivation), similar to the hibernation of hedgehogs in cold climates.

from the heat of the day. They have short tails, well developed eyes and ears (some species have long ears), and long snouts that extend beyond the front of the mouth. They have between 36 and 44 teeth; the front upper incisors have a wide gap between them into which the forward-projecting lower front incisors fit. When a hedgehog closes its mouth, the lower incisors skewer its prey in the gap.

They are nocturnal animals, resting during the daytime in nests within holes in trees or hidden hollows in the ground. Hedgehogs are highly adaptable and will eat almost any creature they can tackle. When they attack adders and other snakes, they keep their spines erect to make it more difficult for the snake to bite their skin. Contrary to popular belief, they are not immune to snakebite.

Woodland hedgehogs living in a temperate climate build nests in which to hibernate during the winter cold. They construct the nests from leaves and grasses or take over old, abandoned nests. At first they may use the nests as temporary refuges for just a few days, but as the weather deteriorates they settle down for prolonged hibernation. Having fed greedily during the warmer months, the hibernating hedgehogs can survive on their reserves of fat. Their body temperature drops to a constant 43°F when they are hibernating, while their heart and breathing rates slow right down. The heartbeat drops from nearly 190 beats per minute to about 20 beats. Such reduction in the body's metabolism decreases the rate at which the fat reserves are used up.

In mild winters, and in areas where the supply of food remains plentiful through the winter, European hedgehogs may not hibernate at all. Hibernation, therefore, is not a fixed habit of the species, but depends on environmental conditions. Indeed, European hedgehogs introduced to New Zealand do not hibernate unless the winter is particularly cold.

Woodland hedgehogs, including the European hedgehog, may breed twice between May and October, before the hibernation season. Tropical hedgehogs do not hibernate and can breed throughout the year, but the desert hedgehogs breed only once, between July and September. Food is scarce all through the year for desert hedgehogs, and

INSECTIVORES AND ELEPHANT SHREWS

the poor food supply is reflected in their litter size. Desert species normally give birth to one or two young, whereas woodland species normally have four or five and sometimes more.

The protective spines of hedgehogs and other mammals provide a good example of convergent evolution at work—the same protective adaptation has evolved independently among several different groups of animals. Tenrecs, which, as insectivores, are distantly related to hedgehogs, possess spines. So too do porcupines and echidnas, neither of which are related to hedgehogs.

When hedgehogs are attacked by predators they roll themselves into spiny balls. So long as European hedgehogs remain tightly curled, one of their main enemies, the fox, cannot get at their vulnerable bellies. However, it is thought that foxes may have learned some cunning ways to outwit hedgehogs. They are reputed to urinate on the hedgehogs and also to push them off grassy banks so that they fall. Apparently these actions startle and confuse the hedgehogs enough to force them to uncurl.

Elephant shrews

Though they resemble shrews in their general body shape and in the way they forage over the ground, the elephant shrews are classified quite separately from the true shrews, and most zoologists now no longer consider them to be true insectivores. Accordingly they are classed in an order of their own. Their name is derived from their long snouts.

Elephant shrews have long legs, large eyes and long tails similar to rats' tails. They live over most of East and southern Africa and part of North Africa, in a variety of habitats, from arid land and open savannah to forests and mountains.

Since they are active during the day instead of during the relative safety of darkness, elephant shrews are shy, secretive animals, always on the lookout for predators. They eat ants and termites, worms, millipedes, crickets, beetles and small amounts of vegetable matter. They forage with their long flexible snouts and catch prey with their long tongues.

Ground dwellers

All species of elephant shrews are strictly terrestrial, spending all their time on the ground. They run quickly on their long legs, and their hind legs are

TOP **The rufous elephant shrew lives in the wooded savannahs of East Africa where its main predators are mongooses, snakes and birds of prey. Elephant shrews are not real shrews and are classified in an order of their own.**

ABOVE **The North African elephant shrew is an isolated species that is found in Morocco and Algeria. The other 14 species of elephant shrews live in a variety of habitats in East and southern Africa, including mountains and forests.**

657

THE EUROPEAN HEDGEHOG
— THE GARDENER'S FRIEND —

Hedgehogs are nocturnal animals that earn their nickname, "gardener's friend," by eating large quantities of the creatures that the gardener regards as pests: slugs, snails, and caterpillars. Hedgehogs are a common sight in both town and country gardens throughout Europe, but sadly they are more commonly seen as the victims of bad driving on the roads. In spite of the continuing deaths of hedgehogs on Europe's highways, however, their population appears to be stable.

European hedgehogs do not live farther north than the north of Scotland and southern Norway and Sweden, since the winters there are too cold. Those in the colder parts of the continent hibernate during winter, living off fat reserves built up in the summer and autumn when food is plentiful. A fully grown hedgehog may almost double its weight during the summer months, laying down thick layers of energy-rich, insulating fat.

European hedgehogs mate at any time between May and October. It is a noisy affair. The male circles the female and brushes against her, making loud snorting sounds. If the female is not interested she butts the male with erected forehead spines. Even if she butts him several times, the male may still persist. When they mate, the female flattens her spines, and the male uses his mouth to grip the spines on her shoulders to keep his balance. They may mate several times before parting.

Born with spines

Gestation lasts between six and seven weeks. Offspring born in the early summer have a plentiful supply of food, and are able to build up enough reserves of fat to provide energy and insulation for the winter. The young are born with their spines beneath the skin to avoid damaging the mother's birth canal. So important are spines for defense, that the white spines that sprout soon after birth change to the brown-and-white spines of an adult's protective coat within 60 hours.

The European hedgehog is primarily a woodland animal, but it has adapted well to gardens and cultivated land. It has strong legs, used for digging the ground in search of food. Apart from

RIGHT A hedgehog feeds on snakes if the opportunity arises. It is not immune to the venomous bite of the adder, but it does show more resistance to it than most mammals. When it preys on snakes, the hedgehog raises its spines as protection against a snakebite, although it is more likely that the snake—not the hedgehog—will be injured in the fight.
BELOW FAR LEFT Although they are ground dwellers, hedgehogs are also good swimmers and will cross small streams and ponds to search for food.
BELOW LEFT Having rolled itself into a spiny, defensive ball, a young hedgehog has been upturned to reveal its soft, wrinkled and vulnerable underparts. By the time it is five or six weeks old, the hedgehog may have more than 2000 spines.
BELOW RIGHT Young hedgehogs cluster around their mother at feeding time. The European hedgehog often has two litters in a year with as many as seven offspring in each litter. She suckles them for at least a month before they fend for themselves.

eating garden pests, hedgehogs also eat birds' eggs and nestlings, mice, lizards, snakes and vegetable matter. Unlike desert hedgehogs, the European species does not dig burrows, but builds a nest for the winter in a sheltered place, taking grasses and twigs, and maybe litter, to keep warm. In cultivated areas it often makes use of the shelter of a log pile.

Feeding hedgehogs

Many people put out food for hedgehogs. The food may be valuable in helping the animals to build up fat reserves in time for winter hibernation, but they are not dependent on it. They will forage over large areas of land and may not visit food bowls for several days, although some individuals do visit them regularly. They may also nest far away from food bowls and do not necessarily follow the same route over their home territories every night. Hedgehogs communicate with one another with loud snorting noises, and they also leave scent trails. Animals of similar size and diet do not usually have such noisy habits.

The hedgehog is able to roll into a ball when alarmed. At first, it may only erect its spines, but if a predator approaches too close, the hedgehog suddenly rolls into a tight ball, almost impossible to penetrate. It tucks its head and legs inside the ball so its vulnerable belly is not exposed. It has a loose, oversize skin to enable it to do this. The tighter a hedgehog rolls itself into a ball, the more erect the spines become.

INSECTIVORES AND ELEPHANT SHREWS

ELEPHANT SHREWS CLASSIFICATION

The 15 species of elephant shrews were formerly classified with the insectivores, but most zoologists now place them in a separate order entirely, the Macroscelidea. They occur over most of East and southern Africa and parts of Central Africa, and in a region bordering the North African coast.

There are four genera. Three species belong to the genus *Rhynchocyon*, including the golden-rumped elephant shrew, *R. chrysopygus*. The genus *Elephantalus* has 10 species, including the rufous elephant shrew, *E. rufescens*, and the North African elephant shrew, *E. rozeti*. The remaining two species have their own genera—the short-eared elephant shrew, *Macroscelides proboscideus*, and the four-toed elephant shrew, *Petrodromus tetradactylus*.

strong enough for them to leap like springhares and jerboas (both rodents). They are all social animals, and there is little difference in the way the various species organize their groups in their different habitats. All elephant shrews have scent glands for marking territory, and some deposit droppings where their territories meet.

Elephant shrews do not dig burrows, but they may use burrows abandoned by other animals or shelter in thick undergrowth. They make a complex network of trails through undergrowth and leaf litter, and these are vital to their survival, since they use them as passages through which they can race away from predators. They defend these trails against intruders and keep them immaculately clean.

Most species of elephant shrews breed at any time and produce several litters each year. Only one or two young are born at a time. They are born with their eyes open and their bodies are covered with hair; in many species the young are able to walk almost immediately. Gestation lasts about 40 days in the larger species and 60 days in smaller species.

Elephant shrews are so secretive that little is known about the status of their populations. Most are not considered to be in any immediate danger, but forest-dwelling elephant shrews are the most vulnerable to loss of habitat.

ABOVE The short-snouted elephant shrew lives in woodland and eats mainly ants and termites. It is active during the day and uses its long legs to make a swift escape from enemies, especially predatory birds. When alarmed, some elephant shrews strike the ground with their hind legs or with their tails. The resulting vibrations serve to warn other elephant shrews nearby that an enemy is in the vicinity. The action is similar to the alarm signals of rabbits and rabbit-eared bandicoots.

INSECTIVORES AND ELEPHANT SHREWS

THE INSECT HUNTERS

The true shrews

Shrews are among the most abundant mammals in the world, making up the fourth-largest family of mammals after the rats and mice, squirrels and common bats. There are more than 240 species in the shrew family. They live all over the world, with a few notable exceptions. There are no shrews in Australia, New Zealand, or New Guinea, none in the polar regions, none in South America south of the Equator and none in the Sahara Desert.

With their long, pointed snouts, short legs and long tails, shrews resemble mice. Their bodies measure less than 8 in. in length, and they weigh no more than 1.3 oz. The short, velvety fur is colored gray or brown. Shrews live mostly on the ground, though some are semi-aquatic. True shrews live in burrows in the ground, in holes among roots, and between rocks or in hollow logs.

Shrews are very nervous animals. The normal rate of their heartbeat is high, but at times of stress it peaks at 1200 beats per minute. The animals are known to die of fright after hearing loud thunderclaps during storms or after being captured by zoologists.

Tropical shrews breed throughout the year; those living in temperate climates normally breed from March to October. The breeding rate is high: shrews have litters of up to 10 young, and females may mate again just one or two days after giving birth. One female may have several litters (in some species as many as 10) in the course of a year.

The gestation period in most shrew species is between two and three weeks, and the offspring are suckled for a similar period. Newborn shrews are hairless and totally blind. Males form no bond with their offspring, but move off in search of another partner after mating.

The size of shrews varies greatly. The largest shrew is the African forest shrew, which may weigh around 1.3 oz. The smallest member of the shrew family, the pygmy white-toothed shrew, is the smallest mammal on earth, weighing less than 0.07 oz. and measuring only 1 to 2 in. from head to tail. Shrews do not appear to have changed much from fossils of their ancestors found from the early part of the Tertiary period, some 50 million years ago. However, they have become slightly smaller.

TOP European common shrews live in woodland, hedgerows, fields and other areas with dense vegetation. They have an enormous appetite and eat several times their own body weight every day. They prey on all kinds of small animals including insects, spiders, wood lice, snails and earthworms.
ABOVE The pygmy white-toothed shrew has uncommonly large ears for a shrew. Though shrews generally have acute hearing, in many species the external part of the ear is barely visible.

INSECTIVORES AND ELEPHANT SHREWS

An armored shrew

One of the most unusual of all shrews is the armored shrew that lives under fallen leaves and in rotting wood in the Central African woodlands. The backbone of the armored shrew has vertebrae that interlock in such a way as to make the spine strong enough to withstand the weight of a large animal. It is possible that the shrew has developed such a resistant backbone in order to prevent it from being crushed by the large herbivores that browse in the woods.

The mole shrew lives in the remote mountains of Taiwan, northern Thailand and southern China. It has a heavy, thick-set body, a short tail, small eyes and short, powerful limbs that are adapted for burrowing underground like a mole (from which it takes its name).

INSECTIVORES CLASSIFICATION: 3

Soricidae

The family Soricidae, the shrews, is by far the largest of the insectivore families. It comprises 22 genera, with a total of 246 species distributed over most of Eurasia, Africa, North America and part of South America.

Nearly half of all shrew species belong to the genus *Crocidura*, found in Africa and Eurasia; the 117 species within the group include the white-toothed shrew, *C. russula*; the African forest shrew, *C. odorata*; and the Egyptian pygmy shrew, *C. religiosa*. The genus *Sorex* contains 52 species, most of which are found in North America and northern Eurasia. They include the European common shrew, *S. araneus*; the European pygmy shrew, *S. minutus*; the least shrew, *S. minutissimus*; and the American water shrew, *S. palustris*.

Ten species of mouse shrew from the genus *Mysosorex* occur in forested parts of Central and southern Africa, and nine species of mountain shrew from the genus *Soriculus* live in China and the Himalayas. The genus *Suncus* contains 15 African species including the pygmy white-toothed shrew, *S. etruscus*, and the house shrew, *S. murinus*. Other species within the family include the Eurasian water shrew, *Neomys fodiens*; the mole shrew, *Anourosorex squamipes*, and the Tibetan water shrew *Nectogale elegans*, of Asia; and the armored shrew *Scutisorex somereni*, of Africa.

TOP AND ABOVE The white-toothed shrew occurs in central and southern Europe and North Africa. When a litter of young white-toothed shrews venture out to explore, they line up behind their mother forming a "caravan." The leading youngster grips the fur at her rump, and the others form a line, each gripping the fur of the animal in front. Mother and young then move off in step; even if their mother is lifted clear of the ground, the young will continue to hang on.

INSECTIVORES AND ELEPHANT SHREWS

Water shrews

Several species of shrews are adapted to a partly aquatic way of life. Although they hunt for food in water, they all forage on land from time to time. The ears of the European water shrew are small and contain valves that are closed when the animal dives. The tail, which has a keel of stiff hairs, is used more as a rudder than for propulsion. The European water shrew feeds on water worms, small fish, frogs and insects. A poisonous saliva, secreted by glands in the lower jaw, is injected into the water shrew's prey.

The Tibetan water shrew is even more highly adapted to aquatic life than the European water shrew. It lives in the mountains of Tibet, Sikkim and southern China at elevations of up to 13,100 ft. Its hind feet are webbed, and its tail is fringed with bristles to give added power in water. It is a skillful swimmer and hunts fish and other aquatic animals.

Moles

There are 29 species of moles and they live in Europe, Asia and North America: they do not occur south of the Equator. They have heavy, thick-set bodies, long, naked snouts, short tails and short, powerful legs. Almost totally blind, their eyes are very small and, in some cases, are hidden under a thin membrane of skin or within the animal's coat. Moles are, however, able to distinguish night from day. The coat is short, soft and velvety and is easy to clean. They are so well adapted to a life of burrowing underground that their forelimbs are sickle-shaped to make efficient digging spades. The hands of the forelimbs are permanently turned outward, and the five fingers on each hand have long, strong claws that are essential for burrowing.

Moles are rarely seen since they live underground almost all the time. They do not have to forage on the surface of the ground but feed on animals that also live in the soil—earthworms and insects, slugs and insect larvae. Moles can move forward or backward

RIGHT Though it frequently hunts on land, the European water shrew is an excellent swimmer and catches most of its food underwater. It has waterproof fur, and its hind toes are fringed with stiff hairs; the toes act as paddles that propel the shrew through the water.

PAGES 664-665 European hedgehogs often have up to seven offspring in one litter. Male hedgehogs do not form permanent bonds with females, leaving them soon after mating.

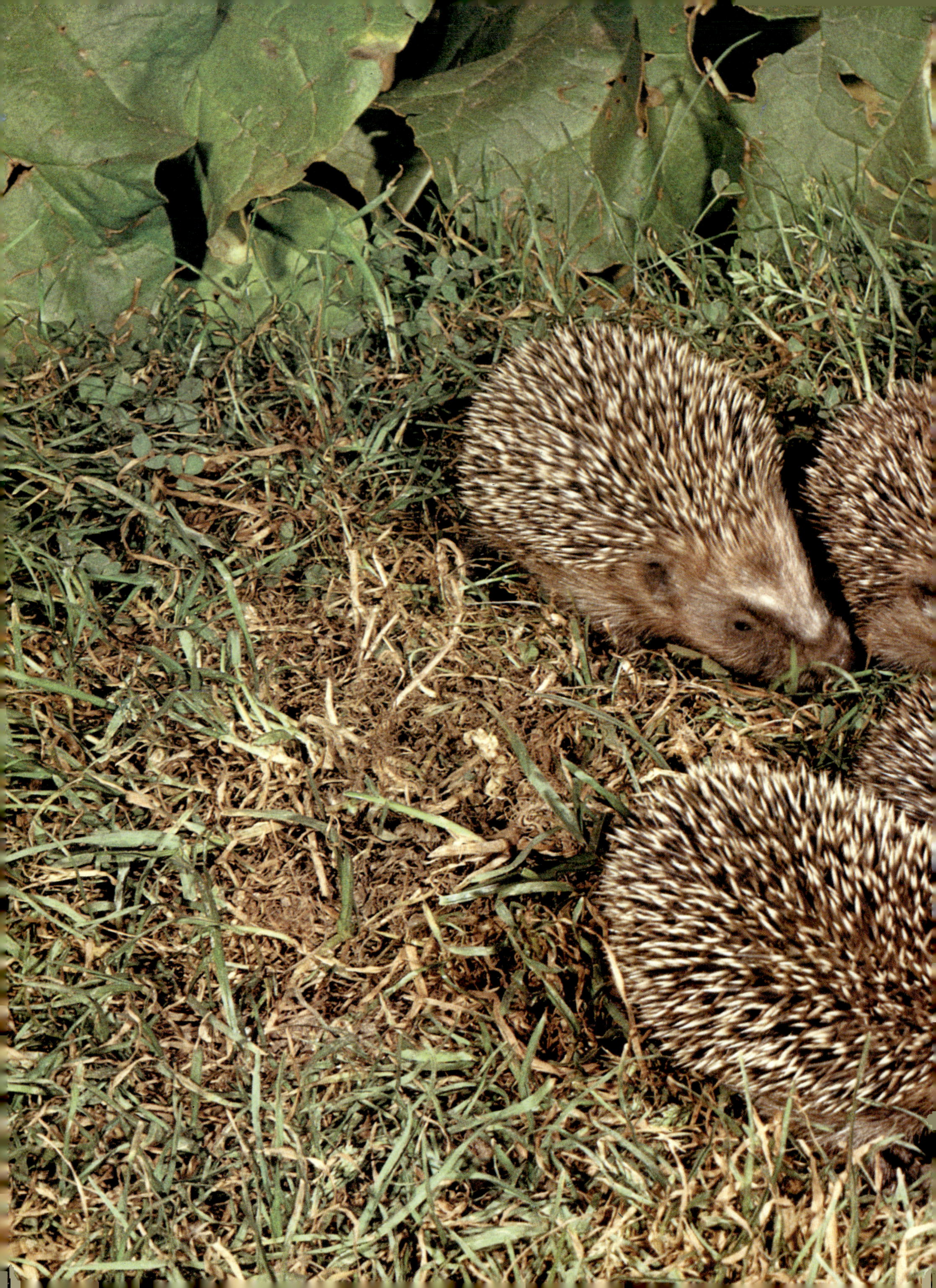

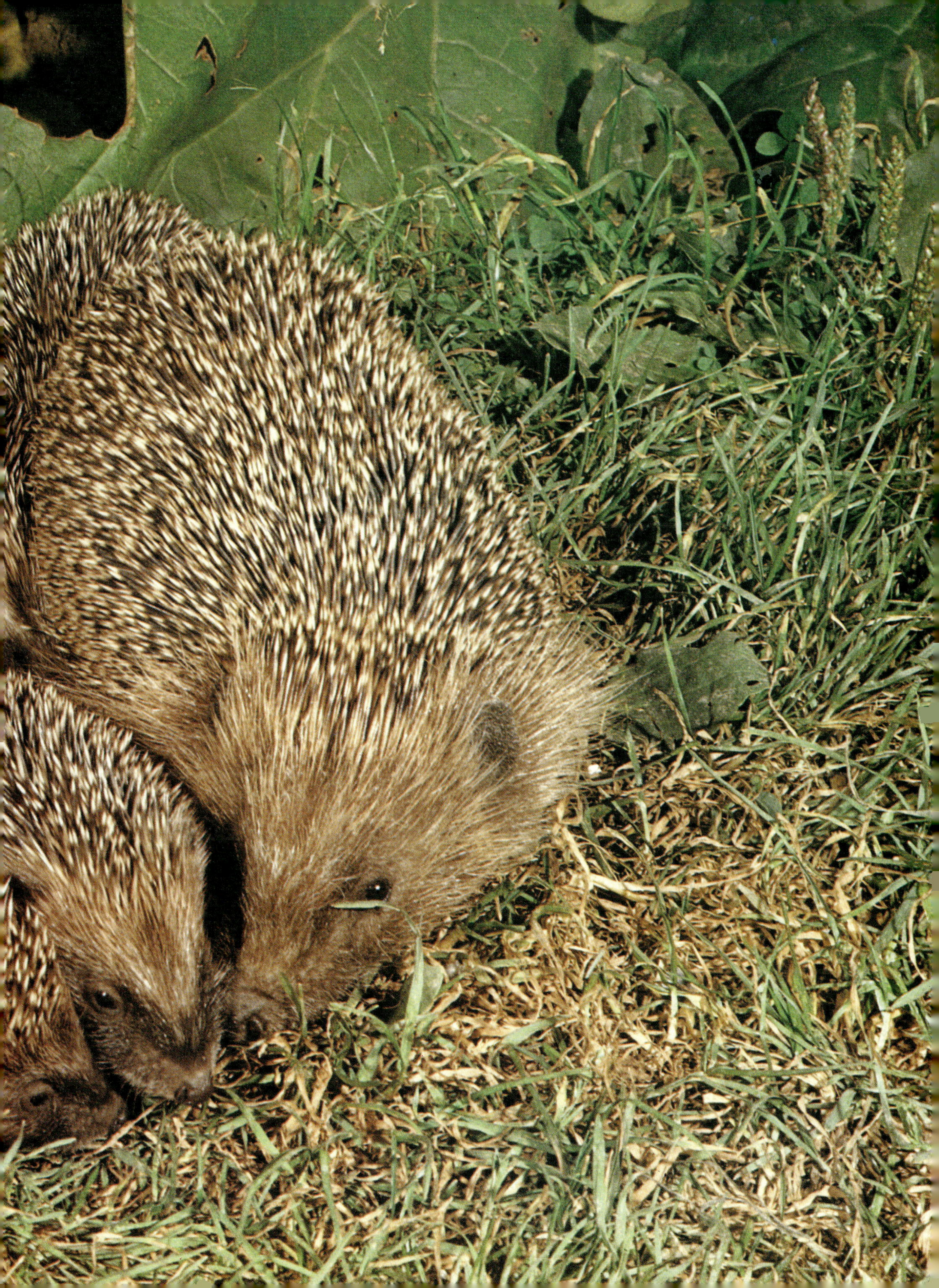

THE EUROPEAN COMMON SHREW
— A FRENZIED SEARCH FOR FOOD —

The appetites of European common shrews—as with most shrews—are so great that the animals are active most of the day and night in their search for food. Each day, European common shrews need to eat their own weight in food in order to survive. They have a very high rate at which body tissues are broken down and built up again, and at which vital chemical processes—the metabolic rate—take place. As a result, they are forced to eat high-energy food every two or three hours; their digestion is so rapid that a full gut can be empty within three hours. The metabolic rate of shrews tends to be higher than that of rodents of the same size, such as mice. Shrews consequently rush around in a frantic search for prey.

A mixed diet

Although they are classed as insectivores, European common shrews feed on a variety of small animals. Using their sensitive snouts and fine whiskers, they forage in leaf litter or just below the surface of the ground, digging out earthworms, insects and insect larvae, spiders, wood lice, beetles and snails. They also eat nuts, berries and fallen fruit. The European common shrew is a tunneler, hiding reserves of food underground so that it can feed undisturbed, out of the way of predatory owls and foxes.

Shrews are born with one set of teeth that lasts their whole life, and it is possible to work out the age of an animal by the amount of wear on the teeth. Teeth wear in older shrews (those aged about a year or eighteen months) can be so extreme that the animals are unable to catch enough food, and they often die of starvation. Younger shrews may also starve to death in winter when insects are inactive, and other creatures are harder to find.

The European common shrew has a habit of licking its own anus. The reason for this behavior is not clear. It has been suggested that in this way the shrew obtains vitamins and trace elements, or bacteria needed to help with the breakdown of food. The shrew is able to reach its rectum by contracting the abdomen, which causes the rectum to turn inside out and protrude from the body.

LEFT The European common shrew is most active at dusk and just after dawn, but since its food requirements are so high, it must also hunt during the day. If the animal goes without food for more than about three hours it will starve to death.
TOP RIGHT European common shrews are highly aggressive to their own kind. When two animals encounter one another they engage in a fierce fight, uttering loud squeaks.
CENTER RIGHT Skin glands on European common shrews secrete a substance that most potential predators find repellent. Domestic cats often kill shrews but they rarely eat them.
BOTTOM RIGHT Even in the coldest winter months, the European common shrew does not hibernate because of the need to eat continually. However, in bad weather it spends more time sheltering in holes in the ground.

Territorial defense

The European common shrews are solitary and defend their territory aggressively. By the time they are eight weeks old they have already established their own territory. If one shrew's territory is invaded by another, the two animals will have a furious, squeaking battle. The tussle may begin with one of the shrews gripping the other's tail in its mouth. The two then grapple with each other, tumbling and spinning on the ground, until the loser breaks free. Fortunately, the fights are largely ritual affairs, and the shrews do not usually harm each other.

When food supplies in an area are low, shrews forage further afield. This may bring them into conflict with other shrews, who aggressively defend their own territories. The story that shrews are dangerous to humans is not true; certain shrews have a poison in their saliva that paralyzes or kills their prey, but will not seriously harm humans. European common shrews only have glands on their flanks, which secrete a smelly substance that repels most animals (except for owls).

INSECTIVORES AND ELEPHANT SHREWS

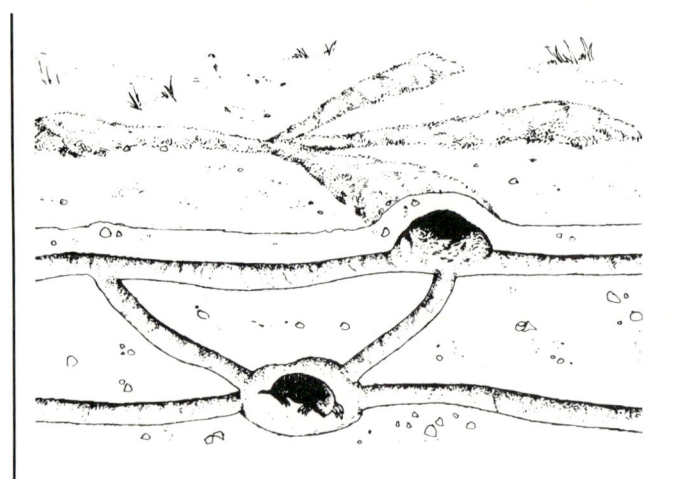

TOP The European water shrew eats mainly small fish, frogs, aquatic worms, and mollusks, but also hunts on land for insects, insect larvae and earthworms. It, in turn, is preyed on by pike, minks and owls.
ABOVE European moles excavate complex networks of tunnels which they defend vigorously from other moles as their hunting territory.
FAR RIGHT A European mole, eating an earthworm, holds the worm down with its claws and wipes away pieces of dirt clinging to the worm's body. European moles feed mainly on the small creatures that fall into their tunnels.

through their tunnel networks—their fur grows at right angles from their skin and can therefore be bent in either direction without hindering their movement. Moles normally live in soft ground, often near rivers and streams. The sickle-shaped forelimbs excavate the soil, which is then thrown back by the hind limbs to keep the tunnel clear.

Molehills

The tunnels of moles may be just a few inches deep or up to three feet deep. The mounds of earth on the surface of the ground—the familiar molehills—are formed when the animal digs a vertical shaft to the surface. The European mole stores food in its tunnel, especially in autumn, to prepare for winter shortage. Moles can eat the equivalent of their own body weight (3-4 oz.) in 24 hours. They are voracious feeders, like shrews, and are active during the day and night.

Moles are solitary, territorial animals. Territories tend to overlap, but neighbors avoid contact as much as possible so as not to come into conflict. They rarely leave their territories, but in times of drought they may travel over half a mile to find water. Moles scent-mark their tunnel walls with secretions from glands in the abdomen. Scent marking has to be done

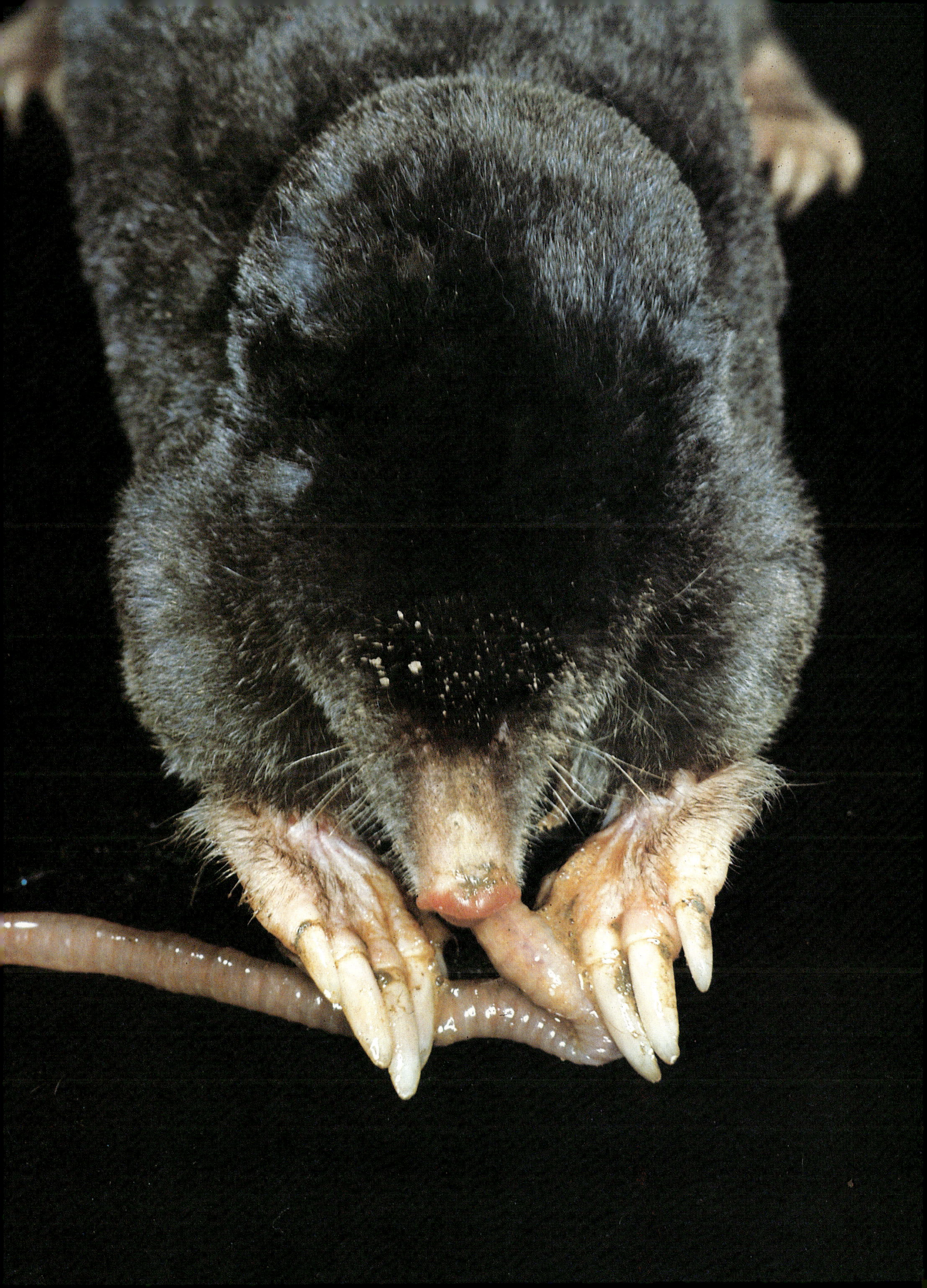

INSECTIVORES AND ELEPHANT SHREWS

INSECTIVORES CLASSIFICATION: 4

Talpidae

The moles and desmans form the family Talpidae. There are 29 species in all, grouped into 12 genera, and distributed over much of Eurasia and North America. The genus *Talpa* is the largest, containing 13 species. Members of the genus include the European mole, *T. europaea*, which occurs in Europe and the USSR; the Mediterranean mole, *T. caeca*, of southern Europe, Turkey and the Caucasus; and the Roman mole, *T. romana*, of Italy and the Balkans. Other species of moles include the star-nosed mole, *Condylura cristatus*, of Canada and the USA and the Chinese or Asiatic shrew mole, *Uropsilus soricipes*, of southern China and northern Burma.

The two species of desmans are the sole members of their genera. The Russian desman, *Desmana moschata*, lives in the lowland rivers of European Russia, while the Pyrenean desman, *Galemys pyrenaicus*, occurs in upland streams in the Pyrenees, northwest Spain and northern Portugal.

TOP A European mole breaks through to the ground surface using its strong forelimbs. Though moles come to the surface to gather nesting material, they are poorly adapted for life above ground. They are slow-moving and vulnerable to predators, and their vision is rudimentary — their tiny eyes can barely distinguish night from day. ABOVE Moles are territorial, marking out their tunnels with scent and fighting aggressively with any intruders. They usually stay within the same territory throughout their lives. FAR RIGHT The newborn offspring of a European mole are blind and hairless at birth (top). When a European mole excavates a new tunnel, it will dig shafts to the surface out of which the excavated soil is pushed, forming a mole hill (bottom).

all the time to prevent the mole's territory from being invaded. In captivity, one mole will not tolerate the presence of another in the same cage. The weaker animal is killed and sometimes eaten by the stronger. In the wild, pairs form only briefly for breeding.

Underground nurseries

In northern and western Europe, moles breed in the spring, and the young are born in the late spring or early summer. The mating season is very short, and it is thought that daylight determines when mating starts. Gestation lasts for about one month and the mother has a litter only once a year, containing between two and seven young. The young are born in nests made of leaves and grasses in special underground chambers. They are born hairless and are suckled for about a month. The males form no bond with their offspring.

The strangest-looking member of the mole family is the star-nosed mole, which has fleshy tentacles surrounding the tip of its snout; these are probably

INSECTIVORES AND ELEPHANT SHREWS

INSECTIVORES AND ELEPHANT SHREWS

delicate organs of touch, which are used to find prey. Males and females of this species live together during winter, which is unusual for moles. Star-nosed moles spend much of their lives in water.

The desmans

The aquatic desmans also belong to the mole family. They have sensitive hairs on their bodies to detect the movement of potential prey. Their tails are flattened and fringed with bristles, which make them broader and more efficient as paddles to steer through water. Their nostrils and ears have valves that can be closed when in water, and they have waterproof fur. Their feet are webbed, and the hind feet are used to propel them through the water. Desmans eat insects and insect larvae, worms and snails, as well as fish and frogs. Unlike most species in the mole family, the desmans are nocturnal.

Moles were once trapped on a large scale in Europe for their fur. Many farmers still regard them as vermin that destroy roots and weaken riverbanks. Consequently they are poisoned irresponsibly, with the result that many species of increasingly rare predatory birds are poisoned in the process. The main threat to desmans is loss of habitat, pollution and damming of rivers.

UNDER THREAT

THE DESMANS

TOP **The Pyrenean desman hunts in fast-flowing mountain streams, feeding on freshwater insects, snails, worms and other small aquatic creatures. Though they belong to the same family as the moles, desmans have adapted to an underwater rather than an underground life. They have long, flattened tails, and webbed feet to provide propulsion through water.**
ABOVE **The Russian desman occurs in pools and slow-moving rivers. It catches larger prey than the Pyrenean desman, often taking small fish and frogs.**

Numbers of Russian desmans began to decrease in the early 1900s when a market was established for their soft, dense fur. Up to this time, they had been fairly numerous in the Don, Volga and Ural river systems, but shooting and trapping rapidly depleted the population. The desmans then faced growing pollution, habitat loss through extensive land drainage, and competition from introduced muskrats and coypu that reduced their food supply and further hampered their chances of survival. Similar problems of pollution and habitat loss have afflicted the Pyrenean desman in its restricted range in the northern mountains of the Iberian Peninsula. Damming of the streams in which the animals live has severely disrupted their environment, and their population continues to decline.

THE COLUGO FAMILY—CYNOCEPHALIDS

TREETOP GLIDERS

The colugos, or flying lemurs, live high in the rain forest canopy where they gracefully glide long distances from tree to tree

THE COLUGO FAMILY—CYNOCEPHALIDS

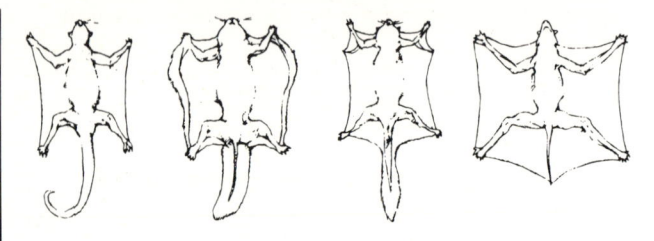

LEFT **A Philippine colugo is displayed with outstretched gliding membrane.**
BELOW LEFT **Gliding membranes compared: sugar glider (A); Siberian flying squirrel (B); scaly tailed flying squirrel (C); and colugo (D).**
RIGHT **A young Malayan colugo lies safe in the hammock-like gliding membrane of its mother, as she hangs from a tree.**
PAGE 673 **A Malayan colugo is slightly larger than its relative from the Philippines, and its grayish-brown coat has more speckles of white fur.**

Colugos, often called flying lemurs, are not in fact lemurs, nor do they fly. They are gliding mammals classified in an order of their own—Dermoptera—meaning "skin-wing," derived from the membrane between their limbs. Recent scientific studies suggest that the colugos' nearest relatives are the primates. There are only two species: the Philippine colugo and the Malayan colugo.

Colugos live in the rain forests and rubber plantations of Southeast Asia; in Thailand, Malaysia, Kampuchea, Vietnam, Sumatra, Java, Borneo and the Philippines. The skin membrane, or patagium, used for gliding, stretches in a broad web from the animal's neck down both sides of the body to the ends of the fingers and toes, and to the tip of the tail. The outstretched membrane acts as a parachute when the animal comes in to land on the tree trunk. It is much larger than the membrane used by other gliding animals (such as the feathertail glider), which only stretches between arms and legs, and not to the tip of the tail. The membrane is covered in hair on both sides.

Colugos are about the size of a domestic cat. Of the two species, the Malayan colugo is slightly larger, measuring 22-27 in. head to tail. Its "wingspan" is around 28 in. and the animal weighs about 4 lbs. The female is slightly larger than the male, and has gray fur with white spots that provide good camouflage when the animal clings to a tree trunk. The male is a more reddish brown color. The coat of the Philippine colugo is darker than that of the Malayan colugo and has fewer white spots.

Master gliders

In the rain forests of Southeast Asia, many animals get their food from within the dense tree cover high above the forest floor. They have devised ways of moving from tree to tree in order to avoid predators on the ground. Squirrels and monkeys overcome the problem with spectacular acrobatic leaps. No animal, though, moves with greater ease between the trees othan the colugos. Their glides are perfectly controlled, and they can cover distances of over 426 ft., losing only about 33 ft. in height over the course. The colugos' limbs are of equal length, giving their patagium an ideal shape for gliding. They have strong, sharp claws for climbing trees and for getting a good grip on tree trunks when landing.

Just before the sun sets, colugos climb to the treetops, stopping frequently for a rest. Before launching into a glide, the colugo turns its head around to survey the forest. Gliding would be impossible in the tangled woodlands of temperate zones, but amid the tall trees of the rain forest there is plenty of open space between the canopy and the forest floor for the colugo to take to the air. When it has chosen which tree to aim for (one that is likely to offer the best leaves for foraging, or a quick escape if the animal is threatened by tree-climbing predators such as snakes), it throws itself into space. Since the glide brings it to land slightly lower down the next tree trunk, it will begin to climb up again for the next launch. Colugos avoid descending to the forest floor as they are slow and clumsy on the ground, and would be defenseless against predators.

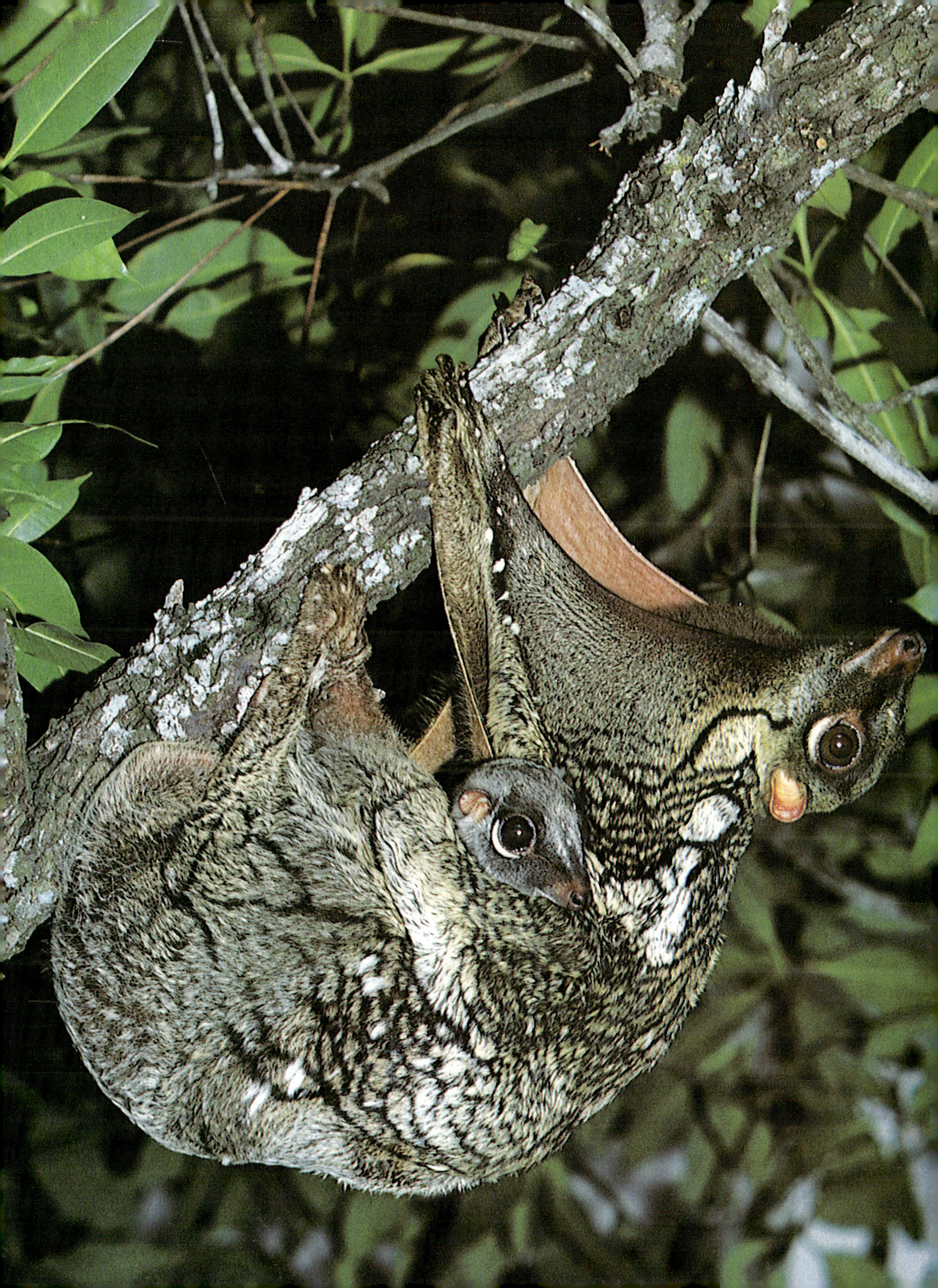

THE COLUGO FAMILY — CYNOCEPHALIDS

ABOVE Colugos have strong claws for gripping trunks and branches, but they are not agile climbers. They will clamber clumsily up to the top of tree trunks before launching into a long glide.

Daytime resting

During the hot, steamy, daylight hours under the rain forest canopy, colugos shelter in the hollows of tree trunks or in the shade of palm fronds on coconut plantations. They are nocturnal animals, and they sleep and feed in separate trees. They often use the same route for gliding through the forest and the same trees for take-off and landing.

Special adaptations

The colugo's large eyes and good eyesight are necessary for judging the distances between trees. As herbivores, they feed mainly on buds, shoots, flowers and leaves. They get all the water they need by licking it from leaves and from the hollows in trees—another way of avoiding the need to descend to ground level. Unlike most mammals, the colugo's large intestine is much longer than the small intestine, and the stomach is extended, enabling the animals to digest the large amounts of leaves that they pull off branches with their strong tongues.

COLUGOS CLASSIFICATION

The two species of colugos or flying lemurs belong to an order of their own, the Dermoptera; their family name is Cynocephalidae. The Malayan colugo, *Cynocephalus variegatus*, is the slightly larger of the two species, and inhabits rain forests and rubber plantations in Southeast Asia and Indonesia. The Philippine colugo, *Cynocephalus volans*, lives in upland and lowland forests on several islands in the Philippines.

Airborne youngsters

Mating occurs between January and March, and gestation lasts for about two months, but a female may become pregnant again before she has weaned her young. Only one offspring is born at a time. Twin births are very rare. As females cannot suckle more than one young at a time, they give birth in quick succession as a means of keeping the population stable. Like marsupials, the offspring of colugos are born undeveloped. The baby colugo is 10 in. long at birth—a third of the length of the mother. The female carries her baby in a pouch, which she creates by folding the patagium just under the tail. She holds the offspring tightly to her chest when she feeds. While the mother forages or glides, the young will grip her teat or belly fur with its teeth to ensure that it does not fall, and it will also hold onto her fur with the claws on its toes for added security. The mother prefers to carry her offspring in the pouch, rather than leave it unattended in the treetops, where it would be highly vulnerable to predators.

Man is the greatest threat to these unique animals—far more so than the Philippine monkey-eating eagles that eat colugos. In rubber and coconut plantations, colugos are shot as crop-damaging pests and they are also hunted by local people for their meat and fur. More importantly, the rain forests of Southeast Asia are being cut down at an alarming rate to clear the land for cultivation or to get at the valuable hardwoods for export to the West. Neither the Malayan colugo nor the Philippine colugo has become an endangered species, but as their habitat disappears, so their numbers will start to dwindle.

THE BATS—CHIROPTERANS

NAVIGATORS OF THE NIGHT

Bats are the only mammals to have mastered true flight. They are mostly secretive creatures, feeding at night and roosting in dark caves, attics and the treetops

THE BATS—CHIROPTERANS

Greater mouse-tailed bat

Indian flying fox

Mexican bulldog bat

Common vampire

Lesser false vampire

Brazilian free-tailed bat

Greater horseshoe bat

THE BATS—CHIROPTERANS

Bats are the only mammals capable of true flight. Other mammals, such as the flying squirrels and colugos (flying lemurs), can glide, often very efficiently, but they cannot perform complex maneuvers in the air or gain height. Bats, by contrast, have developed wings that can be flapped and are as efficient as a bird's wings: many bats are able to fly long distances and have the agility to catch fast-flying prey in the air.

Bats navigate and find food using echolocation—a mechanism that involves sending out high-pitched sounds and picking them up with their highly acute sense of hearing as they bounce off nearby objects. Many night-flying species have limited eyesight, and rely almost entirely on echolocation to orient themselves in the air and identify and capture prey.

The evolution of bats is still a mystery since fossils of primitive bats are rare. The oldest fossil of a bat to be found is very similar to bats of the present day and throws little light on their evolutionary process. The fossil remains—discovered in Wyoming, USA, in what

PAGE 677 **Flying foxes roost in trees during the day, dangling from the branches like strange, leathery fruit.**

BELOW **The distribution of some of the world's bats.** ABOVE RIGHT **A bat's wings are membranes of skin that stretch between its** extended finger bones and its flanks. Fine muscles alter the shape of the wings: during slow flight, the muscles curve the wings to give more lift, but when the bat flies fast, they flatten the wings to decrease air resistance.

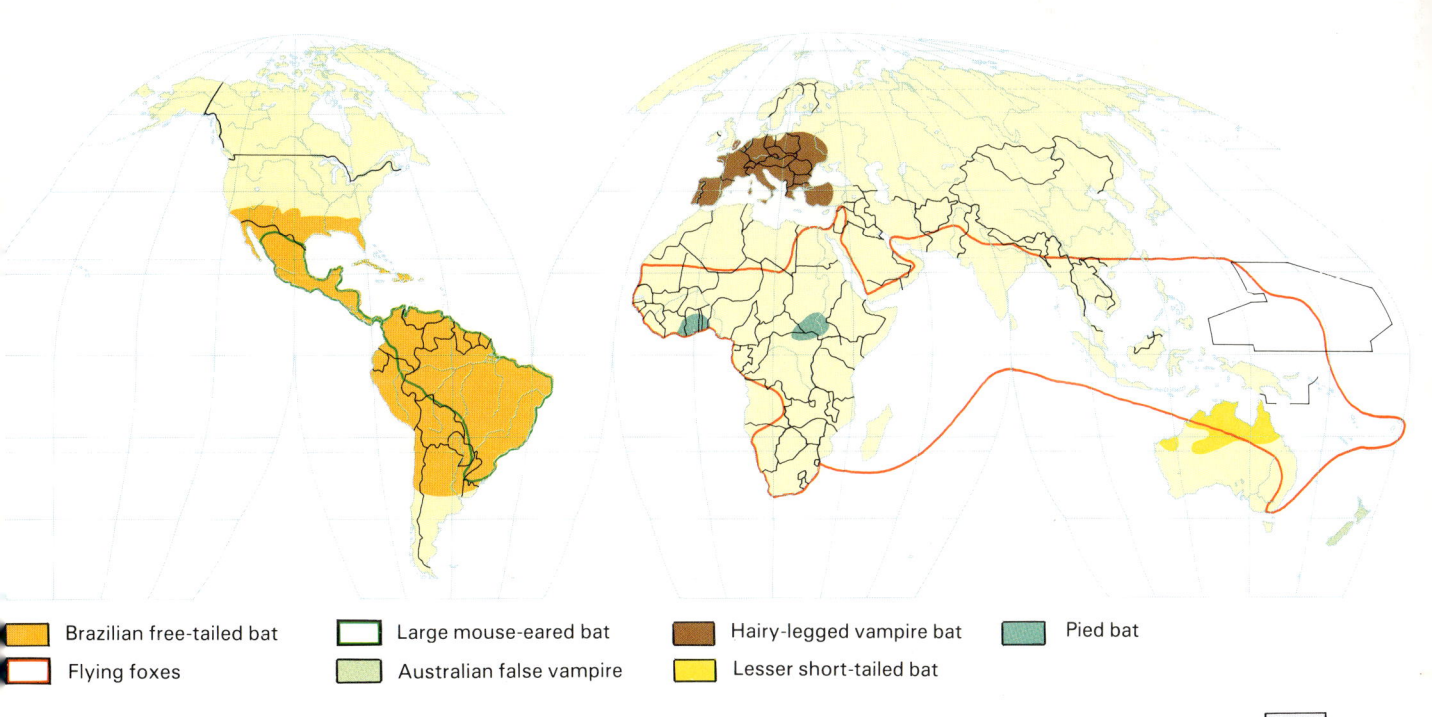

- Brazilian free-tailed bat
- Flying foxes
- Large mouse-eared bat
- Australian false vampire
- Hairy-legged vampire bat
- Lesser short-tailed bat
- Pied bat

679

THE BATS — CHIROPTERANS

Taking to the air

Why did the bats evolve into flying mammals? The reasons are almost certainly to be found in the insectivorous diet of the early bats, coupled with their system of finding food by sound. Many insects can fly, so the air is a good place for an insect eater to find prey. Insects such as moths, flies and beetles have developed a way of keeping an ideal body temperature for flight, even though the temperature of the air may be lower; this enables them to fly at night, escaping most insect-eating birds.

Without competition from the majority of birds, the field was left open to any insect-eating animal that could find and catch flying insects in the dark. The bats, with their echolocation, were ideally placed to take advantage of this. As they evolved, their powers of flight increased, and they could exploit the night-flying insects as their main source of food.

Safety in darkness

Having echolocation and efficient flight made it easy for bats to colonize caves and hollows—habitats that had been neglected by other mammals, except as temporary shelters. Today, they also make their homes in old towers, barns, attics, tunnels, under bridges and inside hollow walls, as well as among rocks and thick vegetation.

As the bats rarely met predators in the caves that they colonized, they could afford to sacrifice their maneuverability on the ground for agility in the air, and so become the specialized aerial hunters that they are today. Their success can be judged by the vast number of bat species—at least 951 (probably with more yet to be discovered)—accounting for nearly a quarter of all mammal species. They are second only to the rodents in size and diversity, and are found in all the continents of the world, except in the polar regions and on the highest mountains.

Vegetarian bats

The primitive bats were all insectivorous, but at an early stage in their evolution one group broke away from the mainstream to develop independently as vegetarians. These bats were to give rise to the flying foxes (also known as fruit bats).

Flying foxes are usually large creatures, with fox or dog-like snouts, and molar teeth that have smooth, flat grinding surfaces for crushing fruit. They also have

ABOVE Its wings wrapped around it like a closed umbrella, a sleeping rousette flying fox shows how a bat's wings are supported by greatly elongated finger bones that form huge webbed hands. By comparison, the hind limbs are short and underdeveloped, although their strong claws give the bat a powerful grip on its perch.

used to be a lake bed—are of a bat that lived about 50 million years ago in the early part of the Eocene epoch. It was called *Icaronycteris* after Icarus—the man in Greek mythology who flew too close to the sun on a pair of man-made wings. It measured about 5 in. head to tail and had fully developed wings (traces of the wing membranes were preserved in the fossil). It may have used echolocation to find its food, probably insects, to judge from the fossilized remains found near to where the bat's stomach would have been.

large eyes, claws on the first two toes of their forelimbs, simple and widely separated ears and short tails with only a narrow tail membrane (or none at all). Flying foxes are tropical, feed almost entirely on fruit, have good eyesight and, usually, a well-developed sense of smell. Only one genus has echolocation skills. Most species neither migrate nor hibernate, although some species journey over great distances to find food. One species that does make seasonal migrations is the straw-colored flying fox of Africa.

The insect eaters

The insect-eating bats developed into a diverse group with 18 families, comprising about 780 species. They eat a wide range of foods, including insects, fish, small mammals, fruit, pollen, nectar and, in the case of the vampire bats, blood.

Generally smaller than the flying foxes, insect-eating bats have short snouts and, in many cases, complex noses with numerous folds and flaps, called leaves, to aid the production of echolocation pulses. Their ears are often equally complicated, with an additional lobe in front of the ear cavity called a tragus. The eyes are small, and although they are not blind, bats navigate largely by echolocation rather than by sight. Many species have well-developed tail membranes. They range farther north than flying foxes, occurring in both tropical and temperate regions. In areas where food is scarce for certain seasons, many species hibernate or migrate.

Bats and flight

All flying animals, be they birds, mammals or insects, have to overcome the same basic problem: since they are heavier than air, they cannot float in it as a fish floats in the water. They have to rely on wings, aided by muscle power, to keep them aloft. The design of the wings is much the same in all these creatures—lightweight wings attached to a light body with the weight concentrated in the middle. Bat wings, like those of birds, are adapted from the forelimbs, but unlike birds, their wings are not feathered. They consist of a double membrane of skin, stretched between the greatly elongated finger bones and extending down to the hind limbs.

The wing membrane, or patagium, consists of many thin muscle fibers sandwiched between layers of skin that allow the wing to be extended or folded. The patagium is rich in nerve endings and is served by a

ABOVE Although highly adapted for flying, the serotine is almost helpless on the ground. It has to drag itself along using its weak, splayed-out hind limbs and the claws on the short "thumbs" that protrude from the wrist of each wing.

BATS CLASSIFICATION: 1

The bats belong to the order Chiroptera. There are over 950 species in all, and they range over all the continents of the world except Antarctica. The order is divided into two suborders—the Megachiroptera and the Microchiroptera. The Megachiroptera is the smaller of the two suborders, containing only one family: the Pteropopidae or flying foxes.

The remaining 18 families make up the suborder Microchiroptera. They are the mouse-tailed bats, the Rhinopomatidae; the sheath-tailed bats, the Emballonuridae; Kitti's hog-nosed bats, the Craseonycteridae; the bulldog bats, the Noctilionidae; the slit-faced bats, the Nycteridae; the false vampire bats, the Megadermatidae; the horseshoe bats, the Rhinolophidae; the leaf-nosed bats, the Hipposideridae; the spear-nosed bats, the Phyllostomatidae; the vampire bats, the Desmodontidae; the sucker-footed bats, the Myzopodidae; the thumbless bats, the Furipteridae; the funnel-eared bats, the Natalidae; the disk-winged bats, the Thyropteridae; the common bats, the Vespertilionidae; the short-tailed bats, the Mystacinidae; the leaf-chinned bats, the Mormoopidae; and the free-tailed bats, the Molossidae.

681

THE BATS—CHIROPTERANS

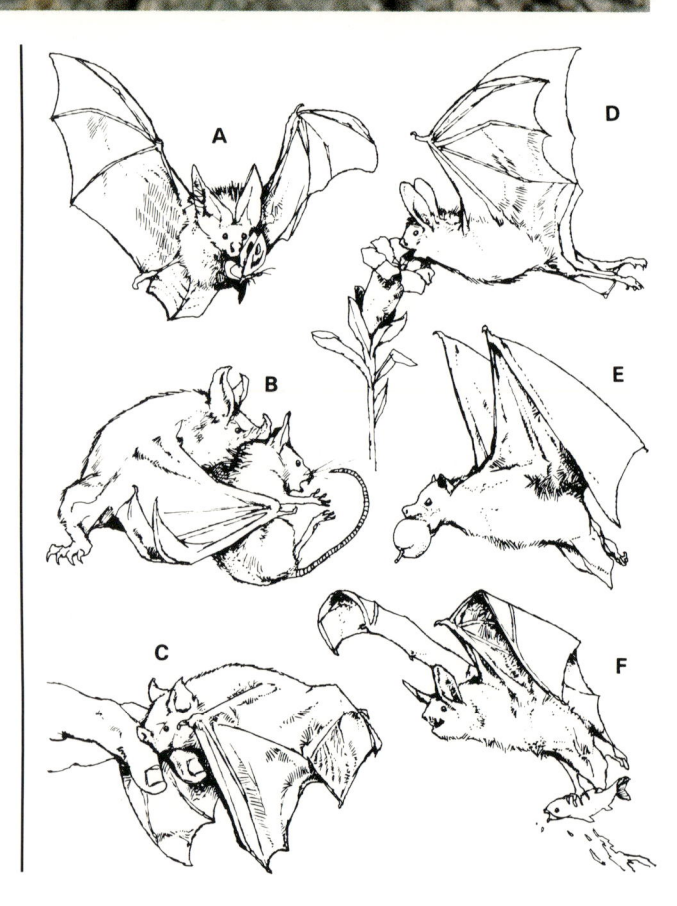

LEFT **A large eyed hawk moth provides a bulky meal for a serotine. Insectivorous bats usually swallow their prey on the wing, but a catch of this size has to be taken to a nearby perch where it will be eaten gradually.**

BELOW LEFT **Bats have evolved a range of diets. The drawings show an insect eater (A), a meat-eating bat (B), a blood-drinking vampire (C), a nectar and pollen feeder (D), a fruit eater (E) and a fish-eating bat (F).**

dense network of blood vessels bringing oxygen to the whole wing area. There are four main flying muscles for each wing. Birds have only two, and both of these are attached to the breastbone, or sternum; as a result, the sternum of a bird has become an important part of its flying equipment and has developed into a pronounced keel. In bats, only one of the four muscles is attached to the sternum; consequently, it has not developed in the same way, and bats are much less deep-chested than birds.

As with birds, the wing of a bat is rounded (convex) on top and hollow (concave) beneath. When an air current meets a wing of this shape, the air current passes in a straight path underneath the wing, but has to travel in a curve over the top. This reduces the pressure above the wing, and since the pressure beneath stays constant, the wing tends to move upward. The lift effect is counteracted by gravity, and the result is a steady glide. By flapping its wings, the bat can also use them to push the air downward and backward, providing the power it needs to gain height and maneuver in the air.

Flight control

A bat is able to exercise very precise control over the movement of its wings; through its tendons and arm muscles, it can alter their shape at any moment, giving the bat its remarkable agility in the air. For example, it can lower the leading edge of each wing and curve the trailing edge down by bending the bones of the fourth and fifth toes; this increases the curve of the wing and gives it more lift at slow speeds, enabling the bat to glide in for the kill before snapping up an insect. Slow-flying species tend to have broader wings than fast fliers, giving them even more scope for altering the wing shape. When a bat beats its wings, most of the movement occurs at the tips; the outer part of each wing (the part farthest from the body) provides most of the forward thrust, while the inner part (nearest the body) produces the most lift.

682

THE BATS—CHIROPTERANS

Most insectivorous bats have small, light bodies to reduce dragging in the air—the lighter the body, the less energy is needed in flight. They have tail membranes to increase the effective wing area and to act as stabilizers and rudders, enabling the bats to dart and weave in pursuit of insect prey. In some species, the tail membrane is used as a means of catching prey. The bat curls it forward to scoop up the insects, and they are either held captive until the bat lands or transferred to the mouth in mid-air.

Feeding on flowers

A number of tropical bat species have abandoned insect eating and have become adapted to feeding on pollen and nectar instead. They generally forage at night, like other bats, often seeking out flowers that grow among the thickest vegetation. They will usually feed without alighting on flowers, but since most of them cannot hover, they must fly very slowly, dipping their tongues into the flowers as they pass by. (Some species have greatly extended tongues that can probe about one inch into flower heads.) Their wings are short, broad, and steeply curved in cross-section to provide sufficient lift at slow speed. The tail membrane is either missing or reduced in size, and as a result, nectar feeders are not as agile in flight as insect eaters.

ABOVE Although they have no feathers, many bats have acquired powers of flight that rival those of birds. A high flier like the noctule bat has long, narrow wings that are ideal for wheeling and gliding in the open sky above the treetops where it catches its prey (mainly cockchafers, moths, and other large insects). Smaller insect eaters hunt low down between trees and buildings, and have shorter, broader wings that allow them to maneuver easily in confined spaces.

Fruit-eating bats, such as the flying foxes, often fly long distances from their roosting sites to their feeding grounds, and are therefore adapted for faster and more direct flight than insect eaters and nectar feeders. Their wings are longer and not as curved in cross-section, allowing them to fly quickly and soar more effectively.

Speed in the air reaches its peak in the free-tailed bats—the mammalian equivalents of the swifts and swallows. Free-tailed bats feed by hunting high-flying insects, and their very fast, soaring flight includes spectacular swoops. They have long, narrow wings with powerful outer sections, and the cross-section of the wing has a very shallow curve, giving minimal lift but also reducing the air resistance. The body is flattened for the same reason, and free-tailed bats have no tail membrane.

THE BATS—CHIROPTERANS

The bats range throughout the world, except in the polar regions and in the highest mountain areas.

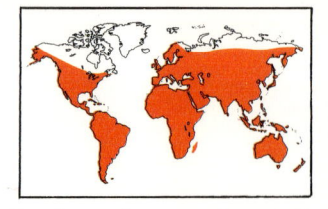

ABOVE In tropical forests, fruit ripens on the trees all year round, and many tropical bats have become specialist fruit eaters. Instead of gnawing at the flesh, they clamp their mouths into it and suck out the sweet juices. Some species have even developed adhesive, sucker-shaped mouths to improve their feeding efficiency.

"Coat hanger" feet

In contrast to its forelimbs, a bat's hind limbs are poorly developed. Their main function is to provide hooks that allow the bat to hang upside down in the roost. Accordingly, the toes are equipped with strong claws for gripping the bark of branches or the cracks in rock faces. The "coat hanger" role of the feet is so effective that researchers are able to remove hibernating bats from their perches, weigh them, and hook them back on without the animals either waking up or weakening their grip. Most bats hang head downward when they are resting or hibernating, although many cave-dwelling species wedge themselves in crevices and grip the rock with the claws on their "thumbs" as well as with those on their hind feet.

Some species of bats can clamber over the ground on all fours, their hind limbs splayed out at 90 degrees to their bodies, giving them a curious posture, reminiscent of insects. Vampire bats, however, are fairly agile on the ground, for they often need to walk over the surface to gain access to their victims. They have short legs and long thumbs to aid their progress, and they can, if necessary, run fast and even jump clear of the ground.

Hunting techniques

Small insectivorous bats forage for prey on the wing, flying through dense vegetation or above ponds and lakes favored by gnats and mosquitoes. In towns, they congregate around the street lamps that attract moths and flying beetles. Each bat tends to fly a regular beat, darting from one direction to another, using its echolocation system to home in on flying insects. Since acute hearing is crucial to their hunting technique, many insect-eating bats have greatly enlarged ears, but because they fly so slowly the aerodynamic effect is minimal.

The system of echolocation that enables bats to find their prey is similar to the sonar or sounding devices used on ships to test the depth of the water. The bat generates sound waves that travel through the air until they hit an object. The waves then rebound and travel back to the bat as an echo. Since the echo from a distant object will take longer to return than the echo from something nearby, the bat can accurately judge how close it is to the object. For us, converting such time delays into distances would involve a lengthy mathematical calculation, but the bat's brain converts the delays instantly to give it a "picture" of its surroundings. The characteristics of an echo can also tell the bat something about the nature of an object: its shape, texture, softness, or hardness. Furthermore, since moving bodies clearly stand out from static background objects, it is easy for a bat to pick out a flying insect.

Unbecoming features

The sophisticated sound generation and reception system of an insect-eating bat has given many species their grotesque faces. The sounds are produced through the nose or mouth, and are often amplified and focused by complex leaf-like structures on the nose (as in the horseshoe bat, leaf-nosed bats and spear-nosed bats). The echoes are received by ears that may have extra convolutions and flaps to pick up every trace of sound. In many species the ears are extremely large to increase their sensitivity.

The sounds bats make in order to find their way and detect food usually take the form of a continuous series of very high-pitched pulses. In horseshoe bats the pulses are normally emitted at the rate of about ten per second when the bats are hunting. The pulses are far too high up the scale for humans to hear; some people may hear bats squeaking, but these sounds are not used for echolocation. Indeed, the echolocation pulses were not detected by scientists until 1938, shortly after the invention of microphones sensitive enough to be able to pick up such high-pitched sounds.

The rapid stream of sound signals enables a bat to monitor the movements of its fast-moving prey with amazing precision. Sometimes less than a second separates the initial detection of an insect and its capture. When not hunting, a bat slows down its sound generation rate; high-flying noctule bats that are simply flying from one place to another produce pulses at a rate of about one per second. The rate rapidly picks up when they begin to hunt, and as a noctule bat closes in on a victim, it may emit up to 200 sound pulses per second.

Moth defense

Though insect-eating bats have evolved such a precise system of hunting, some flying moths have found a way to evade it. The moths are able to sense the sound pulses from nearby bats, and when the predators come too close, they start to zig-zag wildly in the air to confuse the bats and avoid their final swoops. Being smaller than a bat, a moth is more agile, and so if it can judge the bat's line of attack correctly, it can dodge aside at the last moment, just as a matador side-steps a charging bull. Some moths will close their wings and let themselves fall suddenly to the ground under their own weight, dropping away from the bat altogether.

Like other insectivores, such as shrews, insect-eating bats have a high metabolic rate (the rate at which the chemical processes of the body occur), and they need to eat a great deal just to stay alive. In one study, bats caught only half an hour after the start of the night's hunting had already eaten over a quarter of their own body weight in insects. Digestion is also rapid—food can travel the length of the digestive system in a little over half an hour in some species.

Research into the ecology of many insectivorous bat species has revealed the important part that they play

ABOVE True vampire bats feed almost exclusively on the blood of live animals, using their razor-sharp teeth to slit the skin. Contrary to myth, they do not suck the blood but lap it up by darting their tongues into the wound.

PAGES 686-687 A huge flock of flying foxes, or fruit bats, settles down to roost in the treetops. Flying foxes are the largest of all the bats—some species have a wingspan equal in length to that of a large goose.

in the control of insect populations. The huge colonies of Brazilian free-tailed bats, for example, may consume over 6600 tons of insects each year. For this reason, if for no other, the fact that many species of bats are endangered is cause for concern. Dramatic increases in insect populations would be a serious threat to the ecological balance in many areas.

Meat-eating bats

Some bats have evolved the means to tackle prey much larger than insects—the carnivorous bats eat small mammals, birds, lizards, frogs and fish. They are similar to insect eaters but are larger in size. They find their prey using echolocation and also by listening for the sounds made by their prey. The fringe-lipped bat of Central and South America, for example, specializes in hunting male frogs by tuning in to their distinctive mating calls.

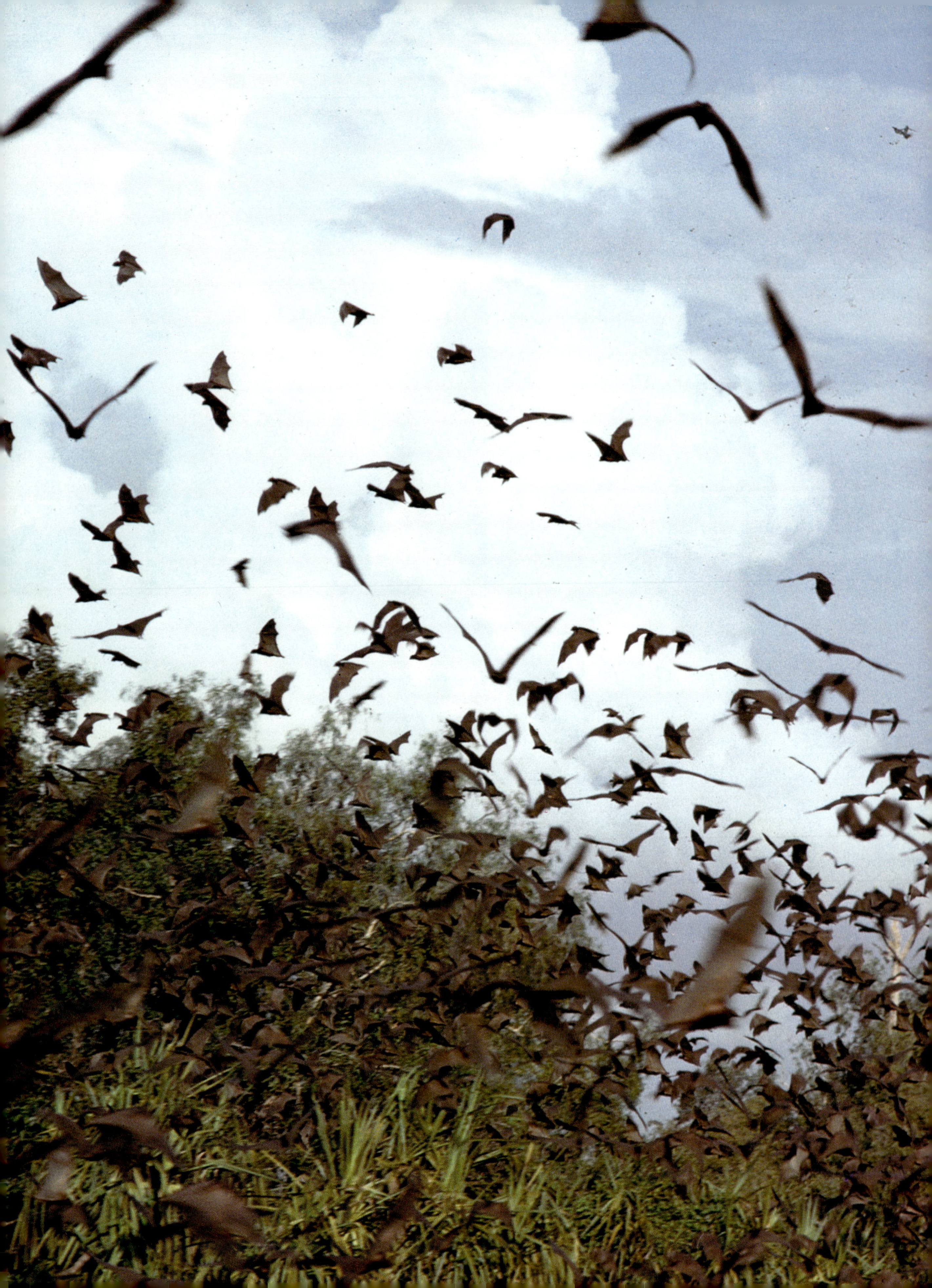

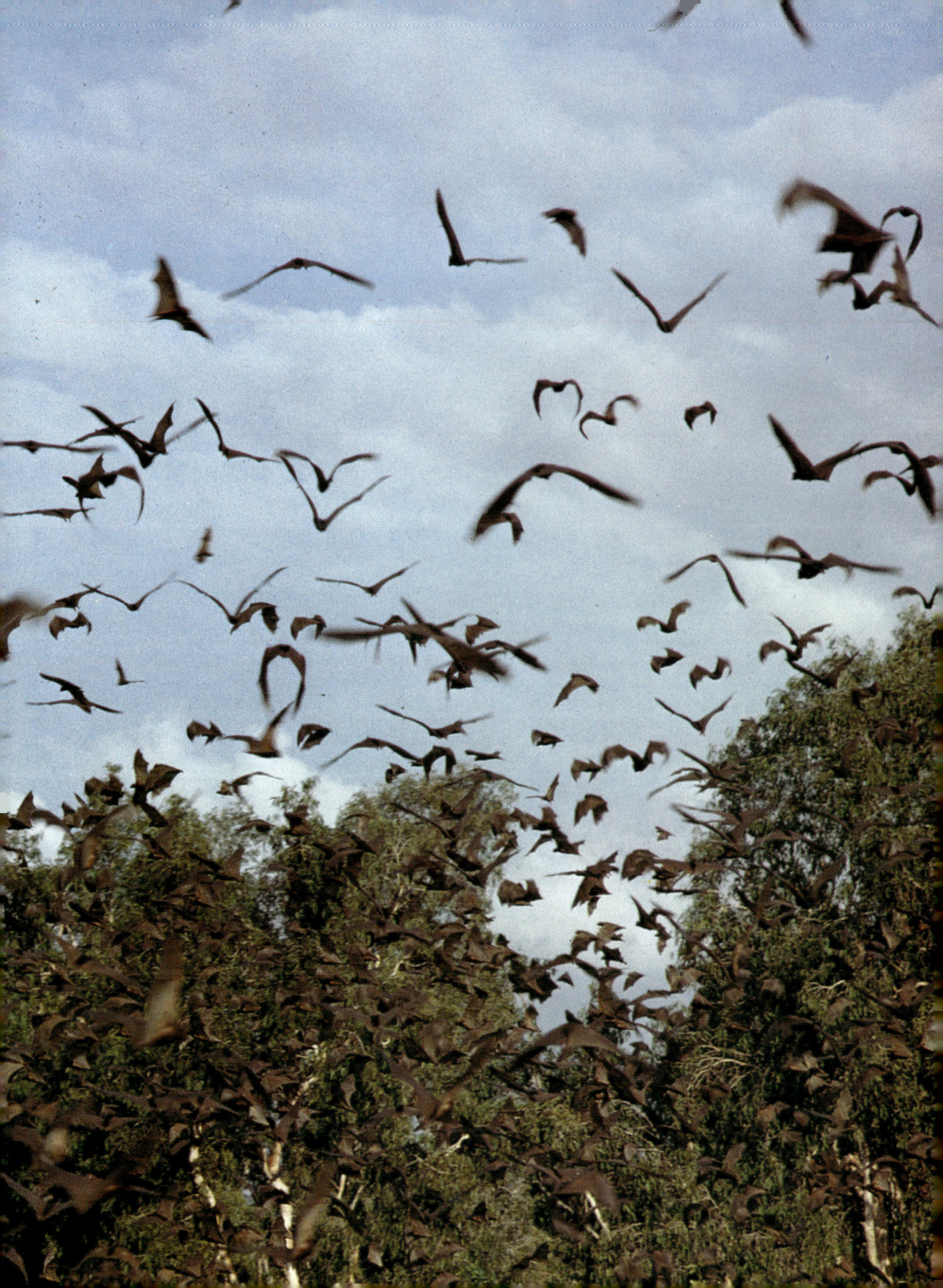

THE BATS—CHIROPTERANS

ABOVE The long-tongued fruit bat uses its hairy tongue for probing into flowers. The tufts of hair act in the same way as an absorbent wick — when the tongue is dipped into the recesses of a flower, the hairs soak up nectar. Some of the pollen from the flower usually sticks to the bat's muzzle and is carried from plant to plant. As a result, the bats contribute to the process of cross-pollination.

THE GRAY BAT

Under Threat

The gray bat of North America is one of the many species of bats throughout the world that are now endangered. Living in large colonies in a small number of limestone caves, gray bats are highly vulnerable to disturbance. They have repeatedly been driven from their roosts by vandals, cavers and even by overenthusiastic biologists. Deprived of a suitable roosting site, many of the bats simply perish, and over three-quarters of their population has disappeared in the last few centuries. Disturbance and habitat destruction are serious threats to all bats, but they have also suffered more than most other animals from deliberate, and unwarranted, persecution.

Milk teeth

Young bats have hook-shaped milk teeth that enable them to hang onto their mother's teats even when she is flying. Eventually, these teeth are replaced by a permanent set that will vary greatly from species to species depending on their diet. Flesh eaters, such as the false vampire bats, have big, sharp, meat-shearing teeth, while at the other extreme the teeth of nectar-feeding bats are reduced to little more than stumps.

Saving energy

Many bats, particularly the common and horseshoe bats of temperate climates, save energy by lowering their body temperature. This is a useful device for an animal that catches insects on the wing, for on days when few flying insects are to be found (in cooler weather, for example), the bat simply stays in the roost and becomes inactive (torpid), allowing its body temperature to drop to a level close to that of its surroundings. Since it no longer has to use up energy in moving and keeping warm, it has no trouble surviving without food until the insects start flying again.

In winter, some bats will hibernate for several months, while others migrate to areas where insects are still to be found. Hibernating species will retire to daytime roosts such as caves, mines or tunnels, but if these are exposed to extreme winter temperatures, they will seek somewhere more sheltered, and may crawl into roof spaces or wall cavities.

Hibernating bats do not suffer in low temperatures. Of all mammals, bats undergo the most dramatic change when they go into hibernation. Their body temperatures may drop below the freezing point of water. In North America, red bats have been found alive and perfectly healthy with body temperatures of 23°F; they do not freeze solid because blood freezes at a lower temperature than water. Such bats often become covered with dew, which does freeze, and so they may spend several weeks encased in ice.

Fat reserves

When in hibernation, bats use only a tiny fraction of the energy they require when active. Whatever energy they do need is supplied from the fat reserves laid down at the end of summer. The fat is stored between the shoulders, and it is here that the highest temperature can be measured when the animal begins to emerge from its torpor.

THE BATS—CHIROPTERANS

Bats are able to enter a state of hibernation much faster than other hibernating mammals of similar size. The first stages are marked by great variations in heart rate, from over 1000 to a few hundred beats per minute. The rate gradually drops lower and lower until it levels off at a point at which the animal enters true hibernation, and its metabolism stabilizes at the minimum level necessary to stay alive. At this point, the heart rate may tick over at only 10 beats per minute. Despite this, hibernating bats still keep some awareness of their surroundings, since they wake up if they are disturbed.

A liking for caves

When bats hibernate, they often hang off the roofs of mines and caves in very large groups. There is still no proper explanation for this behavior; it is often put down to a need to conserve heat, but isolated bats of the same species are, in fact, able to maintain similar body temperatures. Many bats actually prefer to hibernate at very low temperatures and often select cave mouths rather than the warmer areas within. It may be that the dense crowding keeps the temperature of the bats as constant as possible and minimizes the effects of drafts. On the other hand, the animals may hibernate in a group simply because they all congregate in the best place.

Several factors determine which hibernating site the bats choose. Seclusion is important, as well as protection from rain, snow and high winds. It is also essential for the humidity to be high. If it is not, the evaporation of the bats' body moisture means that they have to wake up—an energy-consuming process, since all the body systems have to be started from cold—so that the lost moisture can be replaced by a quick drink. If necessary, bats will obtain water from springs or rock crevices, or even lick off the drops of condensation that often form on their own bodies. However, if they are too active during the winter, they may use up the energy stored in their fat before the insects that they feed on have become plentiful again; and if this happens they will starve.

Non-hibernators

Not all bats hibernate. Many do not need to, since food is available throughout the year. The large tropical flying foxes, for example, can always find enough to eat, and their metabolic rate stays well

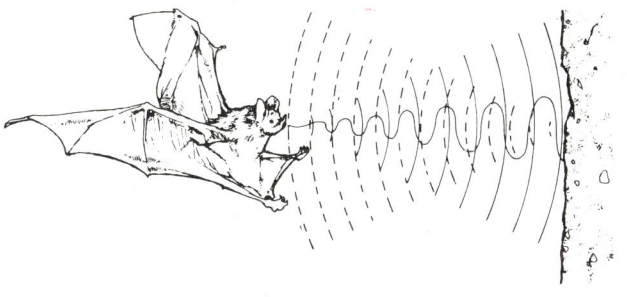

TOP Bats emit high-pitched sounds which they use for echolocation, enabling them to assess their surroundings and locate moving prey in the dark. Though the noctule bat produces the sounds from its open mouth, other bat species make sounds through their noses.
ABOVE By judging the time lag between the emission of a sound and the moment its echo comes back, a bat can accurately measure the distance from an obstacle.

THE BATS—CHIROPTERANS

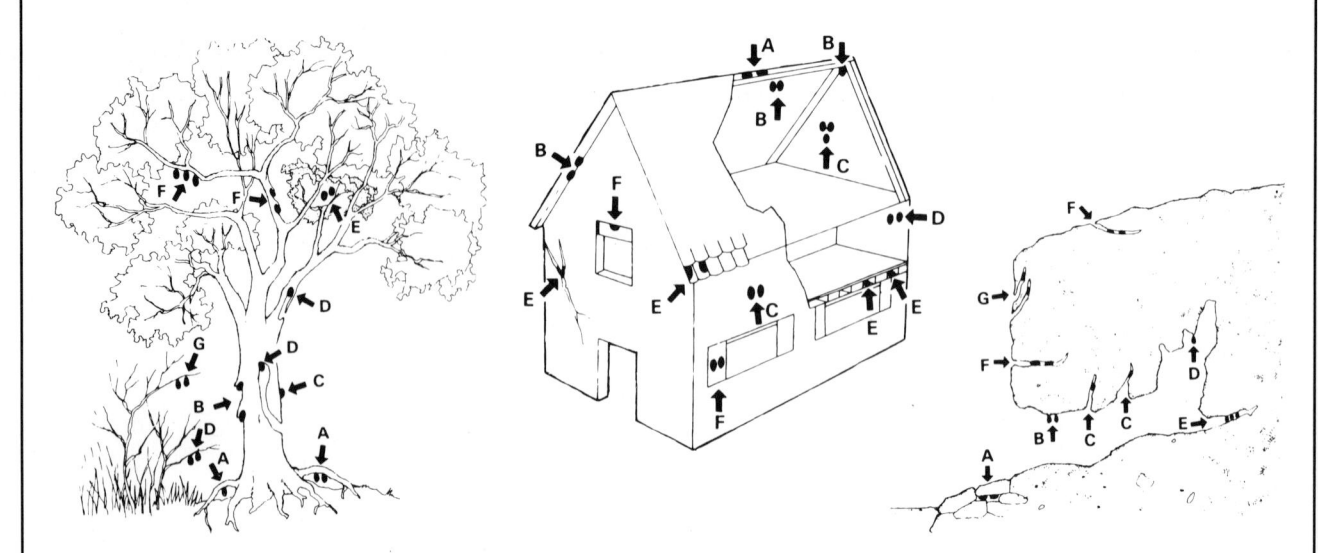

ABOVE All bats have certain preferred roosting sites. The first diagram shows the favorite tree-roosting locations of: epauletted bats—among the roots (A); serotines—under the bark (B); tomb bats—on the trunk of the tree (C); slit-faced bats—in hollow trunks and branches, and in low-lying vegetation (D); red bats—among the leaves (E); flying foxes—beneath the bark and among the highest branches (F); and yellow-winged bats—in bushes (G).
ABOVE CENTER The second diagram shows favored roosting sites in houses: long-eared bats—above the beams (A); horseshoe bats and serotines—under the beams, between walls and the roof, and in cavity walls (B); tomb bats—on the inside walls of barns and outside walls (C); mouse-tailed bats—on inside walls (D); pipistrelles—under floors, tiles and in crevices (E); and barbastelles—behind shutters (F).
ABOVE RIGHT The third diagram shows preferred roosting sites in caves and rock faces: myotis bats—under stones (A); rousettes—in well-lit caves (B); mastiff bats—in cave crevices (C); vampire bats—on roof projections (D); leaf-nosed bats—in horizontal cave crevices (E); free-tailed bats—in horizontal or slanting crevices (F); and tomb bats—in slanting crevices in rock walls (G).

above hibernatory levels, even when they sleep. Indeed, they may even heat up if the surrounding temperature is high; in such situations, the animals spread their wings out and fan them in the wind to allow heat to radiate from the blood capillary networks in the membranes. If the air temperature drops below comfortable levels, flying foxes wrap their wings around themselves like cloaks to provide effective windproofing. Other tropical species may enter a period of torpor similar to hibernation when food is short, but their body temperature remains high throughout.

Great migrators

Some species avoid the need to hibernate in winter by flying off to regions where food is more plentiful. Bat migrations were first suspected when observers noticed that many species of bats that were common in summer disappeared during winter, and were not to be found in any of the likely hibernation sites. The first research into these migrations was carried out in the same way as with birds—by marking individuals with rings on their forelimbs. Later, small ear tags were used, but the principle remained the same: by recording the numbers on the tags whenever bats were recaptured, zoologists were able to build up a picture of each animal's movements.

It became apparent that some species traveled a long distance in search of better conditions. The noctule bat, a large bat equipped for fast, direct flight, regularly travels 560-685 miles to its winter quarters. The migratory journeys of some of the smaller bats are even more remarkable. A common pipistrelle, for example, is recorded as having traveled 930 miles between Russia and Bulgaria. Many species, such as the greater horseshoe bat, migrate in large flocks, although this particular bat rarely travels more than 30 miles. The migration is often little more than a mass decision to find a suitable spot for hibernation.

THE BATS—CHIROPTERANS

Mysteries of navigation

Research has been carried out on bats to find out how they navigate during long migrations, and how they find they way back to their caves from feeding grounds 25 miles or more from the roost. The results have not proved conclusive, and have given rise to several theories. One is that bats use echolocation to recognize certain features of the landscape. Assuming they do not stray over unfamiliar ground, this would enable them to keep a constant check on their position. Ordinary hearing may also help orientation if the roost is situated near a noisy landmark such as a river, road or seashore.

Another theory suggests that sight is an important element of orientation, even for those species that rely heavily on echolocation for hunting and local navigation. Bats that were blindfolded showed that they experienced more difficulty in finding their way back to their colony.

It is still not known exactly how bats use sight to navigate away from their home range. Visual memory alone is surely not enough to explain how long and successful migratory journeys are made by certain species. It has been suggested that bats can navigate by the stars. Incredible as it may seem, this has been shown to be the case with birds; since bats have more sophisticated mammalian brains, there is no reason to suppose they should not do likewise.

Seasonal wind direction is another possible guide for a migrating bat. If the migration takes place at the same time each year, and the prevailing wind always blows from the same quarter in that season, then all the bat has to do is fly at a fixed angle to the wind. This theory has the merit of simplicity, but like the others, it has yet to be proved.

Breeding

Hibernatory species mate at the end of summer. The female stores the semen in her body throughout the winter, and fertilization takes place when she wakes up the following spring. The young cannot develop while she is in her winter torpor. The length of gestation varies, depending on the food supply. If the female becomes torpid again because there is insufficient food to justify hunting, the development of the fetus stops until she becomes active again. One result of this is that the young are always born at times when food is abundant.

TOP While some bats hibernate singly, other species cluster into groups that may become very tightly packed, reaching densities of over 3000 individuals per square yard.
ABOVE Many bat colonies have occupied the same caves for generations, in some cases for thousands of years. Bats are the only mammals to have exploited cave habitats to the full. There, in the absence of predators such as birds of prey, they are able to breed, rest, and hibernate in safety.

THE BATS—CHIROPTERANS

TOP AND TOP RIGHT Shrouded in its wings, the lesser horseshoe bat is well protected from drafts and the evaporation of body moisture. When they are roosting, most species from cool climates allow their body temperature to drop to that of the surrounding air. In doing so, they save energy that would otherwise be used up in the effort to keep warm—a saving that proves particularly valuable when food is in short supply.

The females often give birth in a special nursery area—either a secluded part of the roost or a separate cave. Usually, each breeding female produces one offspring a year. The newborn bat is hairless, blind and helpless, but it already has claws on its thumbs so that it can cling to its mother, even when she is flying. Later, the young bat stays behind at the nesting site while its mother goes hunting, and is suckled at intervals until it has been weaned.

Flying foxes

Flying foxes are found throughout the tropical regions of the Old World from Africa, through Asia to Australia and the Polynesian islands. They usually have thick coats of soft, dark-brown fur, and long, fox-like snouts. Their ears are small and oval, and most of them have no echolocation ability, but their eyes are large and their eyesight is good. Their teeth are adapted for a vegetarian diet consisting mainly of fruit; they have small incisors, unspecialized canines, and broad cheek teeth for grinding vegetable fiber. The first two finger bones on each wing are equipped with claws, used for clinging to tree trunks or vertical walls; some species are also good climbers.

Fruit suckers and pollinators

Flying foxes generally forage at night, recognizing their food by smell. Once they have identified an edible fruit, they hold it between their feet as they drink the juice, leaving the pips and most of the flesh behind. The lips and throat of many species are adapted for sucking fruit juice, and their teeth are rarely used. Other species have long tongues and feed on the abundant pollen and nectar produced by tropical flowers. As they do so, they transfer pollen from one plant to another, and several plants rely on the bats for cross-pollination; in effect, they perform the same function as bees.

Most flying foxes spend the daytime hanging upside down in the treetops. They become active at night, leaving the roost in large groups to fly off to their feeding grounds. The direction they take depends on where the fruit crops are ripe.

THE BATS—CHIROPTERANS

RIGHT A naked bat clings to the wall of a South American cave. When its large wings are folded, the naked bat can tuck its forelimbs into special pockets formed by the wing membranes. The photograph shows one forelimb exposed (on the left) and the other in place within the pocket.

A grooved nose

The short-nosed fruit bat is found in bushy areas in eastern Asia. Four inches long, it has a wingspan of 15 in. and has dark-brown fur. Its stumpy, dog-shaped snout has prominent nostrils separated by a groove, and a divided upper lip. It is a gregarious animal, flying in flocks at night and foraging for the dates, bananas and figs that make up most of its diet. If fruit is hard to come by, it will eat leaves or flowers, particularly those of the banana.

Egyptian rousettes have gained a certain notoriety owing to their habit of roosting in the galleries of tombs, pyramids, and other Egyptian monuments. Easily disturbed, the colonies of these flying foxes frequently take to the air and fly to and fro within the passages, much to the discomfort of Egyptologists investigating the ruins. They are large bats, measuring some 6 in. long with a wingspan of 24 in. Particularly common around the Nile Delta, they range throughout much of Africa and parts of the Middle East and Pakistan.

A bat of the mangroves

The gray-headed flying fox is one of the largest of the fruit bats, with a wingspan measuring up to 47 in. It has a more rounded muzzle than other bats of its family, and although most of its coat is brown, it has a distinctive yellowish neck and shoulders. It lives in the dense mangrove swamps of eastern Australia and forages at night for wild and cultivated fruits such as pawpaws, mangoes and peaches. Like other flying foxes, it depends on its good eyesight and excellent sense of smell to find food. By day, the bat roosts in large groups in the treetops, but despite this, the bats often fall prey to crocodiles, monitor lizards and snakes. They were also traditionally hunted by Aborigines who would light a bonfire under the trees, making the bats drowsy with the smoke, before catching them to eat.

The greater naked-backed bat of the Moluccas, New Guinea and northern Queensland in Australia, is one of the more odd-looking members of the flying fox family. As its name suggests, it has no fur on its back; the wing membranes meet in the middle of its body, along the spine, instead of joining its body at the flanks. This may be an adaptation for flight, possibly giving the bat a greater wing area to help it hover when feeding on fruits. The greater naked-backed bat grows up to 8 in. long and has a comparatively long tail.

The fruit bats are not restricted to lush tropical regions. Some species also occur in arid areas. The Gambian epauletted fruit bat lives mainly in the equatorial forests of Africa where it feeds on fruit, sucking out their juices with its large, fleshy lips. But it is also found on the fringes of the Sahara Desert where it feeds and roosts on the date palms.

Compared with other bats, most flying foxes are elegant creatures with finely drawn faces, but the male hammer-headed bat of Central Africa is an exception.

THE BATS — CHIROPTERANS

ABOVE **Sunlight penetrating the forest canopy glows through the wings of an African epauletted bat, highlighting the blood vessels that supply the membranes. If the temperature rises too high for comfort, tropical tree-roosting bats unfurl their wings to catch the breeze, allowing heat to radiate from the capillary network. They will combat low temperatures by wrapping their wings around their bodies, but unlike temperate species they are unable to tolerate cold weather.**

At 15 oz. it is twice the size of the female. It has a square, swollen snout and very prominent lips. Its mouth is almost entirely filled with air cavities that act as voice resonators, and it has a vast larynx (voice box) that occupies a fifth of the body cavity, extending down beyond the lungs to the diaphragm. All this sound producing equipment is essential to the mating ritual of the bats, which depend upon their loud, metallic-sounding calls to attract females.

The mating buzz

As dusk falls in the dry season, the male hammerheaded bats move from the trees where they roost during the day and congregate in groups of 100 or more in trees usually near a river's edge. These traditional mating sites are visited by the females when they want to select a mate. Each male stakes a claim in the mating tree and starts to beat its wings rhythmically and produce cries of increasing volume at a rate of up to 100 calls per minute.

Attracted by the barrage of noise, the females fly in silently from the surrounding forest and circle slowly over the group of males. If a female is attracted by a particular male, she may settle on a nearby branch, then take off before settling again. She does this repeatedly. Meanwhile, the male speeds up his calls into a fast buzz interspersed with lower cries. Eventually, the female alights next to him and they mate. The pair stays together for less than a minute before the female flies off.

Since all the female hammerheaded bats tend to choose the same group of males to mate with, competition between the males is intense. Those males that can call louder than the others are the most successful and sire most of the next generation.

Like all flying foxes, the hammerheaded bat has relatively small ears and large eyes, and navigates almost exclusively by sight. Males often travel 6 miles or

BATS CLASSIFICATION: 2

Pteropopidae

The flying foxes or fruit bats make up the family Pteropopidae. There are 44 genera containing at least 173 species. These include the short-nosed fruit bat, *Cynopterus sphinx*, of eastern Asia; the Egyptian rousette, *Rousettus aegyptiacus*, which ranges over most of Africa south of the Sahara and from the eastern Mediterranean to Pakistan; the gray-headed flying fox, *Pteropus poliocephalus*, of Tasmania, northern Queensland, New Guinea and nearby islands; the greater naked-backed bat, *Dobsonia moluccensis*, of Australia and the Pacific islands; the hammerheaded bat, *Hypsignathus monstrosus*, of Central Africa; the common long-tailed fruit bat, *Macroglossus minimus*, of Southeast Asia. Borneo and the southern Philippines; Pallas' tube-nosed bat, *Nyctimene cephalotes*, of eastern Indonesia; the straw-colored flying fox, *Eidolon helvum*, of tropical Africa and Arabia; and the Gambian epauletted fruit bat, *Epomophorus gambianus*, of tropical Africa.

THE BATS—CHIROPTERANS

more to find rich feeding grounds, where they spend their nights hanging in the trees feeding on fruit, such as ripe figs. The bats have reduced tail membranes, allowing them to move their hind limbs freely and manipulate the fruit as they eat. They chew the flesh to extract the juice and then let the fruit fall to the ground with its seeds intact; as a result, these bats, like many other fruit bats, ensure that their favorite plants continue to flourish.

Tropical Africa is also the home of the straw-colored flying fox, one of the largest African bats, with a body length of 8 in. It lives in large colonies, feeds on fruit and nectar, and often migrates for considerable distances to take advantage of seasonal food supplies.

A pollen lover

The common long-tailed fruit bat is found in the tropical regions of the Far East from Thailand to Indonesia and northern Australia. It is one of the smallest of the flying foxes and feeds almost entirely on pollen, which it extracts from flowers using its elongated snout and long tongue (this is usually covered with tiny, fleshy projections).

Pallas' tube-nosed bat is another Southeast Asian species with an elongated muzzle. It feeds mainly on

ABOVE Like many flying foxes, Indian flying foxes prefer exposed roosting sites located high in forest trees. They are vegetarians, and rely on their excellent senses of sight and smell to find food. Since flying foxes do not possess powers of echolocation, they do not have any of the strange facial features common to insect-eating bats that are associated with the production and reception of sound waves. These include multileaved noses and large, wrinkled ears.

the juices of various fruits, clinging to the fruit with its claws while sucking at the flesh. Unusual for bats, its coat is not a dull brown color, but a pale yellowish gray with a dark-brown dorsal (back) stripe.

A diversity of bats

The vast majority of bats belong to a much more diverse group than the flying foxes. They have specialized diets of fruit, nectar, insects, small mammals, blood or even fish. However, they do have some features in common. Compared with the fruit bats they have short muzzles, small eyes and large ears, and they are generally smaller. They use echolocation for navigation and hunting, and many species have complex structures on their noses to produce sound, and sensitive ears to pick up the reflected signals. They tend to roost under cover in caves, crevices or buildings.

THE BATS—CHIROPTERANS

Mouse-tailed bats

Bats are not noted for the length of their tails, and many species do without them altogether. The mouse-tailed bats are exceptional, having tails that are as long as their bodies. There are three species in the family, all of them insect eaters. The greater mouse-tailed bat lives in the arid regions of northeast Africa and from the Middle East to India. It is about three inches long, with another three inches of tail that lacks the tail membrane common to many insectivorous species. The bat has grayish fur and large ears, each containing a tragus—an extra lobe in front of the main ear cavity, for improving the reception and identification of sounds.

Like many other insect-eating bats, the greater mouse-tailed bat has to cope with seasonal food shortage. Before winter, it stocks up by accumulating fat in its abdominal region. Later, when insects are scarce, it enters a state of torpor and lets its body cool down to conserve energy. This is not true hibernation, however, for the bat is easily roused from sleep. It has also developed a number of adaptations for survival in arid regions, including the ability to reduce water loss by concentrating its urine.

Tomb bats

The tomb bats of Africa, India and Southeast Asia belong to a tropical family of mainly insectivorous sheath-tailed bats that number over 50 species. Tomb bats have glandular sacs in the tops of their wings that produce a pungent-smelling red liquid. Several species have similar glands in their lower jaws, possibly used for attracting females. Although they use echolocation, they do not have the leaf-like protrusions around their noses like some species (for example, horseshoe bats). The tail is only partly enclosed by the tail membrane, and the very tip protrudes from the upper surface—hence the name "sheath-tailed." They have an agile, dipping flight when they hunt, and spend the daylight

TOP LEFT The lesser dog-faced fruit bat of Southeast Asia is one of the smaller flying foxes, and one of the few bats to depart from the usual gray-brown coloration.
LEFT Of all the flying foxes, the rousettes are the only species that use echolocation to find their way through the dark recesses of caves. The sound pulses they make are of far lower pitch than those of insect-eating bats, and they are clearly audible to humans as a series of metallic-sounding clicks.

THE BATS—CHIROPTERANS

hours in rocky crevices, hollows, old buildings, mines—and tombs, from which their name is derived.

There are 18 species of tomb bats. Some of these form large colonies, while others live in small groups. They all cling to walls with their wings held out at 45 degrees, looking like withered leaves.

The Egyptian tomb bat is probably the most typical member of the family. It ranges from Senegal to India, including northeast Africa, where it is one of the most common of the tomb bats. About 4 in. long, it has brownish fur spotted with white on its back and wings. These bats are powerful fliers, and can often be detected in the air by their continual calling as they capture insects on the wing. They are also agile on the ground and move nimbly among the cracks and crevices of their daytime shelters.

The proboscis, or tufted, bat is a more specialized member of the sheath-tailed bat family. One of the smallest bats, it rarely grows longer than 1.6 inches from head to tail and has a speckled coat of white and gray hairs. It lives in wetlands and near rivers in tropical South America, and captures insects as they fly over the surface of the water. During the day, it roosts on tree trunks that overhang the water.

Hunters of fish

The bulldog bats are natives of Latin America. The family contains only two species, including one of the most interesting of all the bats: the Mexican bulldog bat, or fisherman bat. Unusual for a bat, the males and females are quite different in appearance. The males are brightly colored, with yellow or red-orange fur, while the females are much darker, being a dull brown or grayish color. They measure 4-5 inches long, with pointed snouts and tube-shaped noses with downward-opening nostrils. There are no sound-producing growths on their noses. Their lips are large and drooping, and the upper lip is deeply cleft. The bat has large, pointed ears and long, narrow wings.

TOP RIGHT Like many of the flying foxes, Egyptian rousettes feed on the nectar from tropical flowers. They may travel in great flocks for tens of miles every night from their roosting sites to their preferred feeding grounds.

RIGHT Instead of using echolocation, a flying fox finds its food by using its acute senses of sight and smell. Compared with other bats, it has large eyes for sharp vision and a sensitive nose for identifying the smells of fruit and flowers.

697

THE BATS — CHIROPTERANS

ABOVE Roosting beneath the sheltering fronds of an African palm tree, a group of epauletted fruit bats is protected from heavy rain. All flying foxes roost in the open, often in more exposed places than this, leaving them vulnerable to hunters. Several of the larger species are valued as a source of meat, and, by seeking out a roost, local hunters armed with shotguns can kill whole colonies in a few minutes.

The unusual feature of the Mexican bulldog bat is its feet, which have developed long, sharp claws. In most bats, the hind feet are used for hanging onto rock faces or branches, but in the Mexican bulldog bat they have become powerful hunting tools, perfectly adapted to catching fish. The bat skims over the surface of the water, identifies the position of its prey by echolocation (possibly detecting ripples on the water as fish come to the surface) and then strikes, hooking the fish out of the water with its claws and transferring it to its mouth. The squirming fish is gripped by the bat's fleshy lips and long canine teeth, and then either swallowed in flight or taken back to the roost to be eaten at leisure. In the course of one night, each bat may catch 30-40 fish.

The Mexican bulldog bat sometimes follows flocks of pelicans and stations itself near the bank to pick off any fish that swim in its direction in an attempt to escape the birds. Unlike other bats, it swims well, using its wings like oars. The bat also eats aquatic insects and crustaceans.

The other member of the bulldog bat family is the lesser bulldog bat, which ranges from Honduras in Central America to northern Argentina. Half the size of its fishing relative, it feeds mainly on insects, and its hind feet are relatively unspecialized.

THE BATS—CHIROPTERANS

Slit-faced bats

The slit-faced bats are widely distributed throughout Africa and Arabia, and one species, the Javan slit-faced bat, is found in Southeast Asia. None of the family are very gregarious, and they are among the very few bats with solitary habits, often roosting singly.

The Egyptian slit-faced bat is found throughout much of tropical and North Africa, as well as Arabia and the eastern Mediterranean. It has long fur and large ears. Like all the slit-faced bats, it has a long tail with a curious T-shaped tip that helps support the tail membrane. Slit-faced bats get their name from the distinctive slit that appears to divide their complex, multi-leaved noses in two and that helps to direct the ultrasonic signals used for echolocation. The bats can often be detected in flight by their metallic-sounding cry.

Like many bats, they are primarily insect eaters, but unusually they capture most of their prey on the ground, showing a marked preference for beetles. They also eat scorpions and seem quite undeterred by their venom.

False vampires

The false vampires of Africa, Asia and Australia (the Old World) have little in common with the true vampires and certainly do not drink blood. They are large, fast-flying carnivores, and they often have a startling appearance owing to their multileaved noses. Their large ears are joined at the base, each with a forked tragus (the lobe in front of the opening of the outer ear). They have no upper incisors, but their large canine teeth equip them well for their diet of insects, frogs, lizards, mice and even smaller bats.

The best-known species of false vampire is the Australian ghost bat, so called because of its pale gray fur. It grows to 5 in. or more in length and is a common predator in the wetter regions of Australia. After capturing its prey, it carries it to a favored perch to be dismembered and eaten (the remains are often scattered on the ground beneath) before it flies off again in search of more. Less gregarious than most bats, it lives alone or in small groups.

Horseshoe bats

The horseshoe bats get their name from the horseshoe-shaped flap surrounding the upper lip and nostrils that disperse the ultrasonic pulses used for

ABOVE While resting, a Gambian epauletted fruit bat keeps a wary eye open for predators. Roosting flying foxes face a number of natural enemies, including eagles, large owls, tree-climbing snakes and large lizards.

BELOW If its roost is invaded, an Indian flying fox may clamber toward the intruder—in this case a crow—in an attempt to drive it away. Flying foxes may also threaten each other, screaming and baring their teeth.

699

THE FLYING FOXES
— GIANTS AMONG BATS —

The flying foxes of the tropics are the most dramatic of all the bats—the largest species have wingspans of almost 6 ft. 6 in. With their great leathery wings and their habit of roosting by day in trees stripped of foliage, they are quite different from the small, shy and secretive bats that flutter through the night skies of Europe.

The flying foxes belong to a group of bats that developed independently of the other bats in the early stages of their evolution. Most other bats became predators, relying on echolocation to hunt and navigate. The flying foxes, meanwhile, became fruit eaters, using sight and smell rather than sound to find their food. Because of this, they did not develop the facial growths (nose leaves) that enable other bats to produce and receive echolocation pulses. Instead, they have relatively small ears, but large eyes and sensitive noses.

Wrapped in wings

Flying foxes spend their days asleep in the open, dangling from the branches of favored roosting trees with their bodies enclosed in their wings. When the temperature falls, they wrap their wings more tightly around themselves, often covering their snouts to keep out the cold. On hot days, they hold their wings out, gently fanning themselves and allowing the breeze to cool the blood that courses through the fine capillary networks in the wing membranes.

At sunset the flying foxes begin to stir, then fly off in groups to their feeding grounds. These groups can number several thousand, and the whole flock may fly 60 miles or more in the quest for food.

When they find a good site, they descend on the trees and attack the ripe fruit with a great commotion of cries. The cacophony lasts until dawn, when the flying foxes return to their roost to sleep. They never settle down completely—even when they sleep their grunts and snuffles can be heard at a considerable distance.

In areas where fruit farming is an important part of the local economy, the flying foxes are greatly persecuted since they can ruin whole orchards. Owing to

their large size, they are traditionally hunted by local people for their meat. However, the use of shotguns has meant that the hunting has gotten out of hand, threatening the survival of several species.

The gray-headed flying fox is probably one of the best-known and most typical species. It is an Australian bat, a native of the eastern coastal strip down to Tasmania, where it feeds on a variety of wild and cultivated fruit. About 16 in. long, it has a wingspan of over 3 ft. and an olive-brown coat with a tawny ring around its neck and nape. Its eyes are large and dark, and its ears are small and pointed.

Gregarious by nature, the individual bats seem unable to forage alone or even take the lead in searching for food. At sunset the whole colony will often circle aimlessly in the sky above the roost for two or three hours before setting off in a common direction. They fly slowly and steadily, often at less than 328 ft. above the ground, skimming low over the treetops. None of the species of flying fox are very prolific. The females usually produce one offspring at a time—a low rate of reproduction for a small mammal.

LEFT Colonies of gray-headed flying foxes may strip most of the foliage from their roosting trees to improve the view of the surrounding terrain; an early warning of the approach of predators is vital.
TOP RIGHT A tightly roosting huddle of Egyptian rousettes bare their teeth in alarm. A single cave-dwelling colony of rousettes can contain several million bats.
RIGHT A banana tree in fruit makes a rich prize for a flying fox. Though flying foxes may sometimes damage commercial orchards and plantations, they play an essential role in pollinating crops and dispersing their seeds. In some areas where the animals have been exterminated as fruit-eating pests, commercial production of tropical fruit crops, such as the durian, has actually declined.

THE BATS—CHIROPTERANS

ABOVE **When alarmed by a predator, a colony of straw-colored flying foxes will take to the air, wheeling above the roosting tree until the danger is past. Straw-colored flying foxes regularly migrate in vast flocks of a million or more in order to take advantage of seasonal gluts of fruit. Other flying foxes stay in one area all year round, using a succession of food sources as the various trees come into fruit.**

echolocation. These bats have large ears without a tragus, and short tails that are completely enclosed in the tail membrane. They have large, rounded wings and fly slowly, dipping and gliding to catch flying insects. At the roost, they wrap their wings around their bodies to prevent loss of water, and a sleeping horseshoe bat bears a resemblance to the chrysalis (pupa) of a butterfly.

In all, there are some 69 species of horseshoe bats found throughout the tropical and temperate regions of the Old World, Australia and New Guinea. Most are found in the tropics—only a handful live in the temperate zone. Temperate species hibernate during the winter when food is scarce. They mate in the autumn, but ovulation and fertilization are delayed until spring because the embryos cannot develop inside the mother's body during the period when she is hibernating.

The greater horseshoe bat is a native of central southern Europe, Africa, and Asia, although it also occurs in northern Europe and Britain where it is at the edge of its range and where its status is precarious. It is 4 in. long with gray-brown fur and generally roosts in tight groups hanging from roofs, cave ceilings or inside hollow trees. The females have two false teats that give the offspring something to hang onto when the mothers take flight.

The lesser horseshoe bat is also found in Europe. It is similar to the larger species, apart from its lighter color, and is very sociable, often forming colonies in attics and deserted buildings. Like the greater horseshoe bat, it captures beetles, moths and other insects on the wing by echolocation. It usually hibernates in caves during the winter.

Leaf-nosed bats

Closely related to the horseshoe bats, the 61 species of leaf-nosed bats are distributed throughout the tropical and subtropical regions of the Old World. One of the most characteristic species is the fulvous leaf-nosed bat, found in southern Pakistan, Sri Lanka, India and as far east as Vietnam. It is a small bat with a large tail membrane and a distinctive horseshoe-shaped nose leaf with a trident-shaped appendage. It captures insects in flight using echolocation, and prefers beetles and locusts, often taking them back to the roost to be eaten. It can move its jaw both up and down and from side to side, like a rodent, enabling it to grind up the hard, horny skeletons of its insect prey.

Spear-nosed bats

The large New World family known as the spear-nosed bats includes the biggest of all the American bats, the false vampire of South America—quite distinct from the Old World false vampires. They grow to 5 in. or more in length. Most species have a nose leaf (or skin protrusion) in the shape of a spearhead—from which they take their name. The purpose of the nose leaf is to direct the pulses used for echolocation. Spear-nosed bats also have large, pointed ears.

Most species of spear-nosed bats are insectivores, but several prey on larger animals—including other bats—while some feed on fruit, or pollen and nectar. The flower feeders tend to have pointed noses and long, extensible tongues (they can stick them out a long way to probe into flowers), with a thick cover of bristles that they use to mop up nectar and pollen. As with other vegetarian bats, they probably use smell and sight for orientation, and they play an important part in pollination and seed dispersal. Spear-nosed bats use a wide variety of roosting sites including caves, mines, buildings, trees and abandoned burrows. The more northerly species sometimes migrate for short distances to escape seasonal food shortages.

Despite its name, the Californian leaf-nosed bat does not belong to the leaf-nosed bat family; instead, it is a typical member of the spear-nosed bats. It is about four inches long and grayish or tawny on the back, with long ears and a spear-shaped nose leaf. It feeds mainly on insects that it catches in flight. Less typical is the greater spear-nosed bat of tropical Latin America. It is a large, heavily built bat with a dark-red coat and a glandular sac on its neck that produces an oily secretion. It generally nests in large groups in caves and caverns or in hollow trees, but occasionally it roosts among the foliage of palm trees. A carnivorous species, the greater spear-nosed bat feeds on birds and small mammals, and it will also eat other bats, capturing them in flight by striking them on the head or the back with the claws on its hind legs. If such prey is not abundant, it will eat whatever insects it can find, or even fruit such as figs and bananas.

ABOVE **Pink skin gleaming through its white fur, an albino tomb bat clings tightly to its perch on the wall of a cave. Tomb bats are good climbers, able to scramble nimbly across rock faces and into the crevices of caves, mines and old buildings, where they spend the day. They call continually as they fly, interspersing the high-pitched echolocation pulses with other sounds quite audible to humans.**

Ranging from southern Mexico to southeast Brazil, the tent-building bat has evolved an unusual means of protecting itself while roosting. The bat bites through the central ribs that support the large leaves of many tropical plants. The two halves of a leaf weakened in this way will droop downward, forming a shelter similar to a tent under which the bats can hang. The tent affords them protection from predators as well as from rain, wind and strong sunlight.

Four species of spear-nosed bats, known as flower bats, occur only on the islands of the West Indies. They live in caves and feed on insects and fruit, in addition to the nectar and pollen from flowers. Two of the species, the Jamaican and the Hispaniola flower bats, are very rare, and zoologists have yet to find a live specimen of the Puerto Rican bat—all that we know of the animal comes from the study of its bones, which have been found in caves.

THE BATS—CHIROPTERANS

BATS CLASSIFICATION: 3

Rhinopomatidae, Emballonuridae and Craseonycteridae

The mouse-tailed bats of the family Rhinopomatidae consist of three species in one genus, including the greater mouse-tailed bat, *Rhinopoma microphyllum*, that ranges from West Africa to India and also occurs in Sumatra. The sheath-tailed bats of the family Emballonuridae number over 50 species grouped into 13 genera. They include the Egyptian tomb bat, *Taphozous perforatus*, of much of Africa, the Middle East and India, and the proboscis bat, *Rhynchonycteris naso*, of Latin America. Kitti's hog-nosed bat, *Craseonycteris thonglongyai*, discovered in Thailand in 1973, is the only member of the family Craseonycteridae.

Noctilionidae, Nycteridae and Megadermatidae

The family Noctilionidae of Latin America contains two species in a single genus, one of which, the Mexican bulldog bat or fisherman bat, *Noctilio leporinus*, catches fish. It ranges from Mexico to northern Argentina. The slit-faced bats of the family Nycteridae include 11 species in a single genus. They are insect eaters, but many, such as *Nycteris thebaica* of North Africa, Israel and Arabia, also catch scorpions. The three genera and five species of false vampires that make up the family Megadermatidae include the Australian false vampire or ghost bat, *Macroderma gigas*, of northern and western Australia.

Rhinolophidae and Hipposideridae

There are some 69 species in a single genus within the family Rhinolophidae—the horseshoe bats. They are found throughout the temperate and tropical regions of the Old World. The species include the greater horseshoe bat, *Rhinolophus ferrumequinum*, and the lesser horseshoe bat, *R. hipposideros*, both of Europe, North Africa and parts of Asia. Closely related are the leaf-nosed bats of the family Hipposideridae, with 61 species in nine genera. They include the fulvous leaf-nosed bat, *Hipposideros fulvus*, that lives in Central, southern and Southeast Asia.

ABOVE The lesser bulldog bat is an insect eater that occurs in Central and South America. It usually hunts close to water and sometimes supplements its diet with small fish. Its close relative, the Mexican bulldog bat, preys almost exclusively on fish. Both species track the movements of fish in the water by detecting surface ripples using echolocation. The bat then dives to the surface of the water, lowers its hind feet and stabs its prey with its long claws. The victim is taken to a nearby perch and devoured.

THE BATS—CHIROPTERANS

Hunting like a hawk

Though it sometimes eats fruit, the New World false vampire is a voracious carnivore, with a hunting technique that resembles the behavior of some birds of prey. It perches on a wall or tree trunk, or hangs from the roof of its shelter, carefully scanning its surroundings for small rodents, birds or reptiles. When a potential victim appears, the bat takes off and swoops down for the kill, using its long canine teeth to pierce the skull. It then carries the prey off in its jaws, returning to the perch to eat it. Holding the prey down with the claw of its thumb, it starts to devour it from the head downward, leaving only the tail. The area around a bat perch is often littered with the remains.

Pallas' long-tongued bat lives throughout Latin America from Mexico to northern Argentina. In contrast to the false vampire, it feeds mainly on pollen—although it often eats the insects that it finds in the flowers. It forages by night, flying slowly over the flowers and scooping up the pollen with its long, bristly tongue. It is a small bat, with a wingspan of some 10 in., and lives in large colonies that often take up residence inside buildings and caves.

Smaller still is the yellow-shouldered bat, also found in South America. The males possess tufts of stiff, yellowish hairs sprouting from the front of their shoulders. They have strange snouts with complex nose leaves similar in shape to lily flowers. Their diet consists almost exclusively of fruit.

Another member of the spear-nosed bat family, the Mexican long-nosed bat, is found at high elevations in arid areas. It feeds mainly on the pollen from flowers—especially cactus blooms—and is whitish in color with an elongated nose. It is a migratory, gregarious species, forming large colonies in caves and disused mines. When the bats fly out to feed, they forage in groups, circling plants and swooping down one after another to probe the flowers. When one of them moves

ABOVE LEFT The greater horseshoe bat derives its name from the fold of skin on its face that disperses echolocation pulses.
ABOVE Over the last century, the greater horseshoe bat has declined dramatically. It has suffered from the destruction of its hunting, roosting, and hibernation sites, and has been greatly harmed by the pesticides used to kill woodworm and preserve the timber in roofs. Today bats are under pressure throughout the world, and several species have recently become extinct.
PAGES 706-707 A gliding serotine clearly shows the bone structure of its wings. The "hands" are located halfway across the wing, and the fingers—some of which are longer than the bones of the arm—are splayed out to support the membranes. The two clawed "thumbs" can be seen projecting forward.

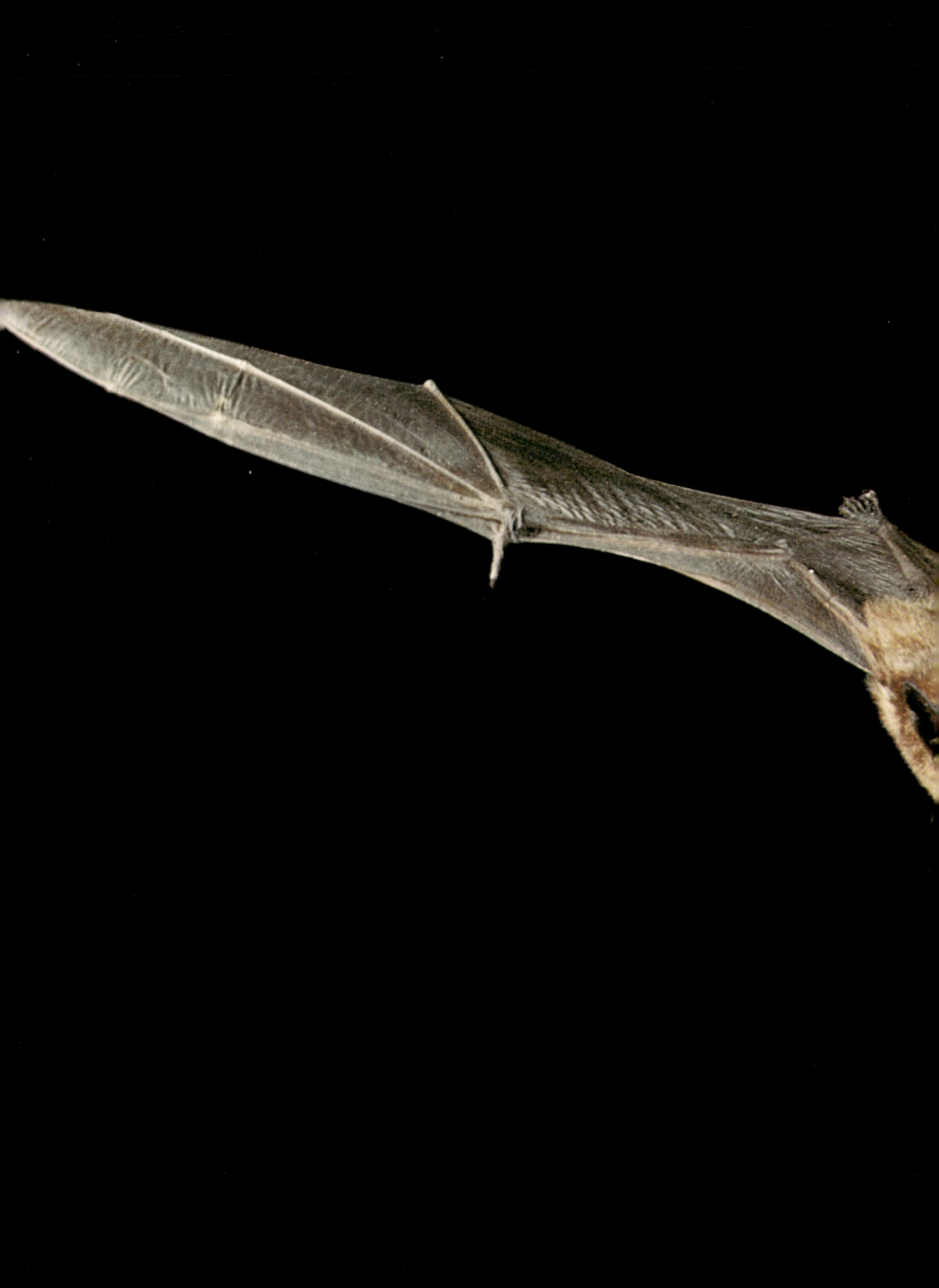

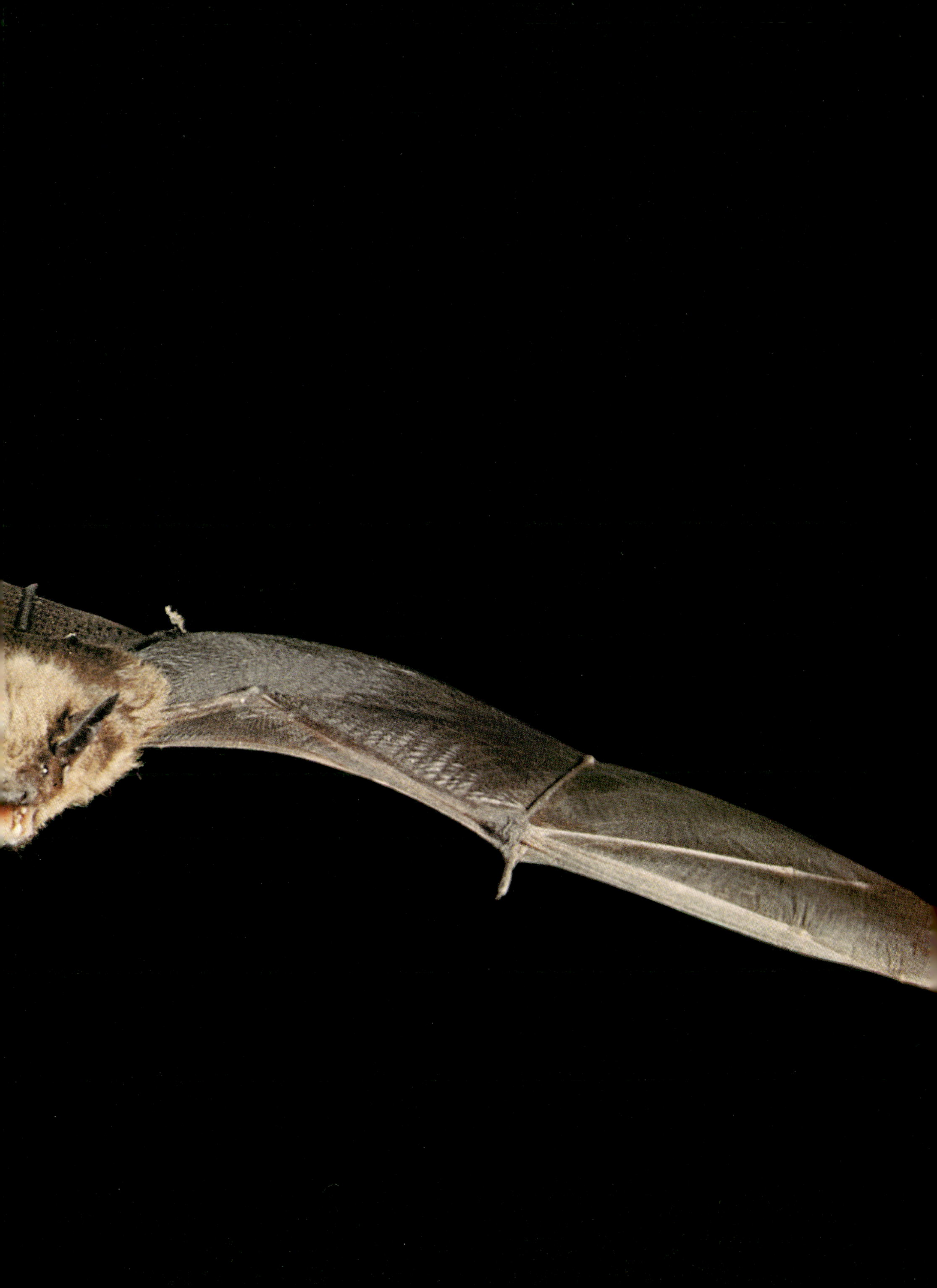

THE BATS—CHIROPTERANS

ABOVE Pallas' long-tongued bat belongs to the New World family of spear-nosed bats. Though spear-nosed bats are primarily insect-eaters, Pallas' long-tongued bat and some other species have adapted to feeding on pollen and nectar, taking over the food source exploited by flying foxes in the Old World.

on to a different plant, the others immediately follow.

One of the strangest looking of the spear-nosed bats is the wrinkle-faced bat of Mexico and Central America. It has a stumpy, hairless nose with much-folded skin. It also produces a strong-smelling secretion from its neck glands. By day it hangs among dense foliage in groups of two or three, emerging at night to feed. An exclusive fruit eater, it has a preference for bananas and papaya, sucking up their juices through a filter formed by its lips and the bristles on its tongue.

Blood-eating bats

The true vampire bats have achieved great notoriety, even though there are only three species. All are found in Central and South America: the common vampire, which usually attacks cattle, and the much rarer white-winged vampire and hairy-legged vampire, which prey on birds. They feed on fresh blood, using their razor-sharp upper canines and incisors to make a slit in the host animal's skin and lapping up the blood as it flows out. Their saliva contains an anticoagulant that stops the blood clotting for as long as the bat keeps drinking.

Vampires generally approach their prey by crawling over the ground, having landed a short distance away. They are much more agile on the ground than other bats, but in appearance they are little different from many quite harmless bats. However, they are quite unlike the large carnivores, now called false vampires, that were once wrongly feared to be bloodsuckers.

Funnel-shaped ears

The two species of thumbless bats are small, South American bats. They have broad, funnel-shaped ears, and their wings are very long in comparison to their bodies. Their thumbs are so small as to be almost invisible. They are insect eaters, but although they have been found roosting in sites such as wine storehouses and a disused sugar mill, zoologists know little about them at present.

Allied to the thumbless bats are the funnel-eared and disk-winged bats. The former are found in Central America; they are lightly built bats with long, slender wings, long hind limbs and very long tails. As their name indicates, they also have large, funnel-shaped ears. There are eight species, and they all prey on small insects that are located and caught using unusually high-pitched echolocation pulses.

The two species of disk-winged bats are natives of the forested regions of Honduras, Brazil, and Peru. They have circular suckers on their thumbs and hind feet to provide a secure grip on the smooth surfaces of large leaves, allowing the bats to roost beneath the leaves under cover from the tropical rain. Like the funnel-eared bats they are insectivorous.

The sucker-footed bat is also adapted to hanging from leaves, and it shelters beneath the fronds of palm trees in its native Madagascar. There is only one species in the family, and it is now very rare.

THE BATS—CHIROPTERANS

RIGHT The large mouse-eared bat is a rare winter visitor to Britain, only recorded occasionally in the counties of Sussex and Dorset. Further south in Europe it is still common in many places.
BELOW RIGHT A Geoffroy's long-nosed bat displays its sharp teeth, used for grasping insects in flight.

BATS CLASSIFICATION: 4

Phyllostomatidae

The family Phyllostomatidae contains some 47 genera and 140 species of spear-nosed bats. They include the Californian leaf-nosed bat, *Macrotus californicus*, of the southwestern USA and northwest Mexico; the greater spear-nosed bat, *Phyllostomus hastatus*, of Central and South America; and the false vampire bat, *Vampyrum spectrum*, which ranges from Mexico south to Brazil and Peru. Other species include Pallas' long-tongued bat, *Glossophaga soricina*, of Central and South America; the yellow-shouldered bat, *Sturnira lilium*, of Central and South America and the Caribbean; the Mexican long-nosed bat, *Leptonycteris nivalis*, of Mexico and Guatemala; and the wrinkle-faced bat, *Centurio senex*, of Central America.

Desmodontidae and Myzopodidae

The true vampire bats belong to the family Desmodontidae. There are three species in three genera: the common vampire, *Desmodus rotundus*; the white-winged vampire, *Diaemus youngi*; and the hairy-legged vampire, *Diphylla ecaudata*. All are natives of Central and South America. The Myzopodidae contains only one rare and little-known species: the sucker-footed bat, *Myzopoda aurita*, of Madagascar.

Furipteridae, Natalidae and Thyropteridae

The family Furipteridae, the thumbless bats, consists of two species in two genera, including the smoky bat, *Amorphochilus schnablii*, of western South America. Two similar families are the funnel-eared bats of the Central American family Natalidae, with nine species; and the two species of disk-winged bats in the family Thyropteridae, including the Honduran disk-winged bat, *Thyroptera discifera*.

THE COMMON VAMPIRE
— NOURISHED BY BLOOD —

The common vampire of Central and South America has been shrouded in mystery and legend for centuries, and its habit of drinking blood from its sleeping victims has probably done more than anything else to give bats an unjustifiably unpopular image.

Vampire bats feed exclusively on fresh blood that they obtain from live mammals and birds. Their teeth are highly specialized, with chisel-like upper incisors and large, razor-sharp upper canines. The teeth are used to make a small incision in the victim's skin, enabling the vampire to drink the blood as it flows out. The bat's saliva contains substances that stop the blood from coagulating; so as long as it keeps drinking, the blood will keep flowing. Once the bat has drunk its fill and flown away, the blood will clot— assuming another vampire is not waiting to take its turn. Each bat may consume up to 40 percent of its own weight during one feed. It has been calculated that a vampire bat will drink some 21 gallons of blood during its life if it survives for the average life span of 13 years.

In the midnight hour

Most vampire activity takes place on the darkest nights, for it is at these times that their victims are least active. A vampire stands little chance of getting a meal from a large grazing animal if it is on its feet. Moreover, it runs a serious risk of being crushed. An animal lying down is a much safer target, and if it is sleeping it may not even feel the bite, delivered as it is by teeth which are as sharp as a scalpel.

Once a victim has been located, probably by smell or by sight, the vampire bat lands on the ground nearby and climbs up the animal's fur until it finds a bare patch of skin. The amount of blood taken by the bat—or even by several successive bats—is insignificant to a large mammal. Nevertheless, a vampire attack can prove fatal. The bats are carriers of the deadly rabies virus, more commonly associated with dogs, and since the virus is transmitted through the saliva, vampires are one of the easiest means by which the disease is spread. A vampire bat can be infected for some time before it finally dies; meanwhile it may have passed the virus on to a different animal every day.

There have been many attempts to calculate the percentage of vampires that are infected at any one time, by catching the bats and counting the

numbers that show traces of the virus. However, since diseased bats are much easier to catch than healthy specimens, it has proved difficult to make an accurate assessment of the figure.

Cattle followers

The rabies threat posed by the common vampire is of great concern to farmers who live in the regions where the bat occurs, since it preys almost exclusively on domestic livestock, and cattle in particular. Indeed, the present-day distribution of the bat reflects the density of cattle farming: the bigger the cattle population, the more vampires there are and the greater the rabies threat. Some estimates put the number of yearly cattle deaths from the disease at about two million.

Occasionally, vampire bats feed from humans, usually biting them on the cheeks or the toes. Most bites are recorded among people who frequently sleep outside or live in poorly protected shelters. On Trinidad in the West Indies, a survey revealed that about 20-30 bites had occurred in three weeks among all children aged under 15 years old. In some areas of Guyana in South America, some 50-70 percent of the population have been bitten at least once during their lifetime.

LEFT Despite their large, complex ears vampires do not rely on echolocation to find their victims. They depend instead on their eyes and on their keen sense of smell.
TOP RIGHT Part of a vampire colony at their roost in a Central American cave. Seasonal flooding of caves frequently forces vampire colonies to migrate.
CENTER RIGHT Unusually nimble on the ground, a vampire often lands at a safe distance from its victim and hops toward it with its wings folded.
RIGHT The common vampire usually seeks out domestic animals for its night-time meals of blood. The other two species — the white-winged and hairy-legged vampires — prefer to feed on the blood of birds.

THE BATS—CHIROPTERANS

TOP **The common pipistrelle is Europe's smallest bat, and one of the most numerous. With a body reaching only 2 in. in length, but with a wingspan of up to 10 in., it was once known as the flittermouse.**

ABOVE **Savi's pipistrelle occurs in mountainous areas of Europe. Like other bats, its teeth are quite primitive and unspecialized. It is probable that the ground-dwelling ancestors of the bats had very similar dentition.**

The adaptable vesper bats

About a third of all bats are classified as vesper or common bats. They are distributed worldwide, and occur in both tropical and temperate regions where they have colonized the whole spectrum of habitats from arid semideserts to lush forests. They are generally small, uniform in color (although some have white spots and patches), and are equipped with scent glands that produce a distinctive smell. They have small eyes, ears that range from the small to the very large, and—in most cases—simple muzzles without leaf-like sound radiators.

Most species are cave dwellers, although many have taken to roosting in tunnels, attics, farm buildings and hollow trees. They are mostly insect eaters, but several species will prey on small mammals, reptiles, amphibians and fish. In temperate regions, where insects are scarce in winter, many species either migrate to areas where insects are abundant or conserve their energy by hibernating.

The largest European

The large mouse-eared bat is widely distributed from southern Europe to the Middle East, and occurs sporadically farther north. It is the biggest of the European bats, weighing up to 1.6 oz., with a head-body length of 3 in. and a wingspan of up to 15 in. It has a light, gray-brown coat, whitish underside and large, pointed ears.

BATS CLASSIFICATION: 5

Vespertilionidae

The largest bat family is the Vespertilionidae—the vesper or common bats. There are 319 species (and probably more yet to be discovered) in 42 genera, including the large mouse-eared bat, *Myotis myotis*, ranging from southwest Europe to Syria; Bechstein's bat of Europe and the western USSR; the common pipistrelle, *Pipistrellus pipistrellus*; the noctule, *Nyctalus noctula*, and the serotine, *Eptesicus serotinus*, of Europe, North Africa, and southern Asia; the particolored bat, *Vespertilio murinus*, of Scandinavia, eastern Europe, and southwest Asia; the pied bat, *Glauconycteris superba*, of Central Africa; and the western barbastelle, *Barbastellus barbastellus*, of Europe and North Africa. Other species include the common or brown long-eared bat, *Plecotus auritus*, of Europe and Asia; Schreiber's bent-winged bat, *Miniopterus schreibersi*, of Africa, southwest Europe, southern Asia, and Australia; the bronze tube-nosed bat, *Murina aenea*, of Malaysia; and the pallid bat, *Antrozous pallidus*, of North and Central America.

THE BATS—CHIROPTERANS

Despite its long wings, the large mouse-eared bat has a slow, rather labored flight, yet it is known to make migrations of 124 miles or more between its summer and winter quarters in northern parts of its range. It colonizes a wide range of habitats, preferring open woodland or farmland, but it also occurs in towns where it roosts in cellars and attics during the day. It takes to the air after sunset, hunting on the wing for beetles and moths. It is often attracted by the scores of insects that gather over stretches of open water or in the light cast around street lamps.

Like many of the vesper bats, large mouse-eared bats hibernate in winter. The bats gather in large groups in the cool, sheltered conditions of a cave or tunnel, often traveling some distance to find a good site. The first to emerge in spring are the older animals, including many of the females that have mated the previous autumn. Once their mothers begin to hunt for food again, the young start to develop in the womb. In due course, the females withdraw to secluded nursery roosts where they give birth to single offspring; these become self-sufficient after a few months.

The common pipistrelle

One of the best known of all the vesper bats is the common pipistrelle. It is the smallest of the European bats and also one of the most common, occurring in both town and country. It often lives in large colonies of a thousand or more, and frequently roosts in buildings. Like other bats, it is most active at night, emerging just before sunset to hunt for insects on the wing. It will also sometimes hunt by daylight, particularly on warm, sunny winter days. But for most of the winter it hibernates, preferring cool, dry sites such as attics or barns.

In contrast to the pipistrelle, which hunts close to the ground, the noctule is a high flier. One of the largest European bats, its long, flat wings (with a span of 14 in.) give it a powerful, direct flight pattern. It preys on flying insects, catching them high above the

TOP RIGHT Breeding male noctule bats occupy fixed roosting sites, to which they attract females by making repeated flights of display and uttering loud calls.

RIGHT The common pipistrelle hunts for moths and flies near ground level, quartering the areas between buildings or trees. It, in turn, is hunted by predators such as owls.

THE BATS—CHIROPTERANS

ABOVE LEFT With its fearsome jaws, the particolored bat of eastern Europe and Asia is well equipped to deal with its prey—large flying beetles.
ABOVE A long-eared bat peers down from its roost, looking more like a moth than a mammal. The elongated ears are highly sensitive receivers, designed to collect the rapid stream of echolocation signals that are bounced back from the bat's surroundings.
LEFT When giving birth, a female bat temporarily abandons her upside-down roosting position and folds her tail membrane up to provide an instant cradle for the newborn young.

ground, and on summer evenings it will often join swifts and swallows in their aerial hunting. Noctules live in all types of habitat, although they favor areas with plenty of woodland since they prefer to roost and hibernate in hollow trees. However, they also nest in attics, church towers and barns, and are often seen in built-up areas. They have short, rounded ears and broad, dark muzzles with numerous scent glands that give the golden-brown fur a strong smell similar to that of garlic.

Superficially similar to the noctule, the serotine is another large European species. It is a strong flier, but unlike the noctule it tends to hunt at low level, diving down between trees and buildings to catch large insects such as flying beetles and big moths. It prefers flat areas around woods and near riverbanks, but is also found in hilly country. Large colonies often nest in attics, returning to the same site year after year for breeding but hibernating elsewhere in hollow trees or deserted buildings.

White-tipped fur

The particolored bat is common in eastern Europe, and is particularly aggressive toward other bat species. Each hair in its coat is tipped with white, giving it a speckled appearance. It is a fast flier and does most of its hunting around the edges of woods and tree-lined avenues, preying on large insects such as cockchafers. Its wings make a characteristic beating sound as it flies. It generally nests under rocks, but it may also colonize urban habitats and its range is becoming more extensive as towns grow in size. It migrates in autumn to its winter feeding grounds and hibernation sites. In the USSR, such migrations may cover 525 miles or more.

Though not abundant, the barbastelle has a wide distribution throughout Europe, West Africa and North Africa, particularly in mountainous areas. It has a short muzzle and thick ears with reinforcing folds at the front that meet between its tiny eyes. Its long, stiff, blackish fur becomes pale at the tips in older animals, giving them a "frosted" look. About 2 in. long with a small head, short tail, and narrow wings, it has a quick, agile flight and hunts insects in open woodland and river valleys. It generally nests in hollow trees and hibernates in caves during the winter months.

Stripes and warts

The pied bat is found throughout Central Africa and is unusual among bats because its fine tawny coat has well-defined stripes and its nose is covered in warty growths. It lives in the rain forests, roosting on palm fronds or on wooden sheds during the day and hunting for insects by night with a fluttering flight similar to that of a butterfly.

The five species of long-eared bats are probably the most easily recognized of all the bats, owing to their enormous ears. At about one inch long in the common long-eared bat of Eurasia, the ears are almost as long as the animal's body. Common long-eared bats hold their ears either erect or parallel to the ground during flight, to pick up and decode the "sound-picture" reflected in sound waves from their surroundings and potential prey. They fly relatively slowly, fluttering in zigzag fashion on short, broad wings, searching for insects among trees. Aided by their giant ears, their echolocation system is extraordinarily sensitive. They catch their prey on the wing, often scooping up their victims in a pouch formed by folding the tail membrane up toward the belly. Alternatively, they may simply pluck creatures from leaves or twigs.

At rest, the common long-eared bat's ears droop and are hidden under the wings. Normally found in woodland, they roost in hollow trees, often alone. Though they sometimes hibernate in the same location, they usually fly off to find a more sheltered spot such as a cave or mine tunnel.

A wide-ranging bat

Schreiber's bent-winged bat has one of the widest distributions of all the vesper bats. It frequents hot, sparsely populated regions from southern Europe, through Africa and southern Asia, to Australia. Like the noctule, it has long, narrow wings, a swift, direct flight pattern, and it hunts insects at high elevations. It has small ears and a long tail that is completely enclosed by the tail membrane.

TOP The tropical long-fingered bats live in parts of Europe, Asia, Australia, and Africa. They are fast-flying predators, hunting high above the ground.
ABOVE Free-tailed bats are swift fliers. One species, the Brazilian free-tailed bat, lives in huge colonies that may catch over 278 tons of insects each night. When bats fly in dense groups, they utter sounds of slightly different pitches, and listen out for the echoes corresponding to their own calls, so as not to confuse each other's echolocation system.

THE BATS—CHIROPTERANS

Vesper bats also occur in the New World. The pallid bat of North and Central America has an unusual diet, consisting not only of insects but also of scorpions, amphibians and even small lizards, which it catches by skimming low over the ground. It has a sand-colored coat and a nasal flap or leaf in the shape of a horseshoe. It roosts and hibernates in large groups in caves, rock crevices and hollow trees. The females form groups up to 100 strong in April, to give birth and nurse their young.

The Indiana bat lives in the eastern USA, from New Hampshire south to Florida. It roosts colonially in caves and mines during the winter, but is highly specialized in its choice of roosting sites. The air temperature in the cave should be no more than four to eight degrees above freezing. Since few caves are so cool, the bats tend to become concentrated at a limited number of sites. It has been estimated that over 90 percent of the population inhabits the 10 most popular roosts.

BATS CLASSIFICATION: 6

Mystacinidae and Mormoopidae

The family Mystacinidae contains one genus and two species: the greater short-tailed bat, *Mystacina robusta*, and the lesser short-tailed bat, *M. tuberculata*, both of which live in New Zealand. The Mormoopidae or leaf-chinned bats are a slightly larger family, with two genera and eight species including the Antillean ghost-faced bat, *Mormoops blainvillei*, of the Greater Antilles and the Bahamas.

Molossidae

The Molossidae or free-tailed bats comprise 91 species in 12 genera. Of these, 50 species belong to the genus *Tadarida*, including the European free-tailed bat, *Tadarida teniotis*, which ranges from southern Europe and North Africa to the Far East; and the Brazilian free-tailed bat, *T. brasiliensis*, which occurs from the USA south to Argentina. The other species include the naked or hairless bat, *Cheiromeles torquatus*, of Southeast Asia, Indonesia, and the Philippines.

Walkers and climbers

The two species of short-tailed bats form a distinct family and occur only in New Zealand. The most striking characteristic of these forest-dwelling bats is that they can both climb and walk well. They have long, sharp claws on both hind feet and on each thumb. When folded, their wing membranes are enclosed in pouches that prevent accidental damage. They have long snouts and well-separated ears, each with a large, pointed tragus (the lobe in front of the main ear opening).

Free-tailed bats

Most of the free-tailed bats are found in the tropics. The family derives its name from the fact that the tails are only partly enclosed in the tail membrane. The bats are heavily built, with large, broad heads and large, forward-facing ears. They have well-developed hind feet that enable them to climb well. Their long, narrow wings are adapted for fast, direct flight, and they generally fly at high levels with their mouths held open, finding and catching beetles by echolocation. They nest in caves, in old buildings or under rocks.

The European free-tailed bat is the only species of free-tailed bat to be found in Europe. Particularly common in Mediterranean areas, it is about 5 in. long with patches around the throat, ears, and head. It has a large, blunt snout, long ears and a short tail, and it preys mainly on moths.

By the million

The Brazilian (or Mexican) free-tailed bat occurs in both North and South America from the southern USA to Brazil. It forms larger colonies than any other mammal or bird in existence. The size of some congregations is staggering—nursery colonies numbering 50 million bats have been recorded in the southern USA. Most of the very large colonies nest in caves, and when such colonies take to the air, they funnel out of the cave entrances like dark plumes of smoke.

The strangest member of the family of free-tailed bats is the naked bat of Southeast Asia. It is the largest of all the bats, excluding the flying foxes, and is completely hairless apart from its head and the base of its neck. On its neck, a few short hairs adorn a throat sac that produces a strong-smelling secretion. The naked skin is thick and black, giving the bat a tough, leathery appearance. Using the well-developed claws on its front limbs, it is an excellent climber.